P9-BJT-994

LET THERE BE WATER

LET THERE BE WATER

ISRAEL'S SOLUTION FOR A WATER-STARVED WORLD

SETH M. SIEGEL

THOMAS DUNNE BOOKS ST. MARTIN'S PRESS NEW YORK

THOMAS DUNNE BOOKS.
An imprint of St. Martin's Press.

Printed in the United States of America. For information, address St. Martin's Press, 175 Fifth Avenue, New York, N.Y. 10010.

www.thomasdunnebooks.com
www.stmartins.com

Designed by Steven Seighman

Library of Congress Cataloging-in-Publication Data

Siegel, Seth M., 1953–
Let there be water : Israel's solution for a water-starved world / Seth M. Siegel.
pages cm
Includes bibliographical references and index.
ISBN 978-1-250-07395-2 (hardcover)
ISBN 978-1-4668-8544-8 (e-book)
1. Water-supply—Israel. 2. Water resources development—Israel.
I. Title.
TD313.I75S54 2015
333.910095694—dc23

2015025201

First Edition: September 2015

10 9 8 7 6 5 4 3 2 1

For Rachel Ringler,
wife, best friend, partner, and inspiration

As a face opposite water reflects another face,
so do people reflect each other's hearts.
PROVERBS 27:19

Contents

Timeline

1920—The British Mandate of Palestine begins, covering the territory that today is Israel, the West Bank, and Gaza.

1937—Mekorot, which grows to be Israel's national water utility, is founded.

1938—Piped water is brought to the Jezreel Valley, south of Nazareth. The large-scale water infrastructure project is the first in the Land of Israel in modern times.

May 1939—The British White Paper is issued, severely limiting Jewish immigration to Palestine. British Mandate officials make the first of several claims that, due to inadequate water resources, Mandatory Palestine must restrain population growth.

July 1939—In response to the White Paper, the Zionists develop a national water plan showing sophistication in integrated water resource planning and management.

1947—Through deep drilling, water is found in the Negev desert, a source of irrigation for new desert farms.

May 14, 1948—The British Mandate for Palestine ends. The State of Israel is declared.

July 1955—The Yarkon-Negev Pipeline opens, bringing water from the center of Israel to desert farms in the south.

August 1959—A comprehensive water law is passed giving the Israeli government control of all water sources and usage. A powerful regulatory body, the Israel Water Commission, is created.

June 1–2, 1964—President Lyndon Johnson and Prime Minister Levi Eshkol meet and discuss desalination at length during the first Israel state visit to the US.

June 10, 1964—The National Water Carrier opens creating a national water system.

1966—Drip irrigation equipment is offered for sale for the first time.

1969—The Shafdan wastewater treatment plant opens.

1989—A pipeline for treated water from the Shafdan facility to Negev farms is opened.

1995—The Palestinian Water Authority is established as part of the Oslo II agreement between Israel and the Palestinian Authority.

2000—Dual-flush toilets are made compulsory in all new installations in Israel.

2005–2016—Five large seawater desalination plants are built along the Mediterranean coast providing a majority of Israel's drinking water.

2006—The Israel Water Authority, a technocratic, apolitical successor to the Israel Water Commission with broad powers, is created.

2010—Water pricing at real cost begins throughout Israel. Municipal water utilities are established, removing control of water and sewage from Israel's mayors.

October 2013—The Israeli government declares water independence from weather.

December 2013—The Red Sea–Dead Sea agreement is announced by Israel, Jordan, and the Palestinian Authority.

March 2014—An Israel–California water cooperation agreement is announced.

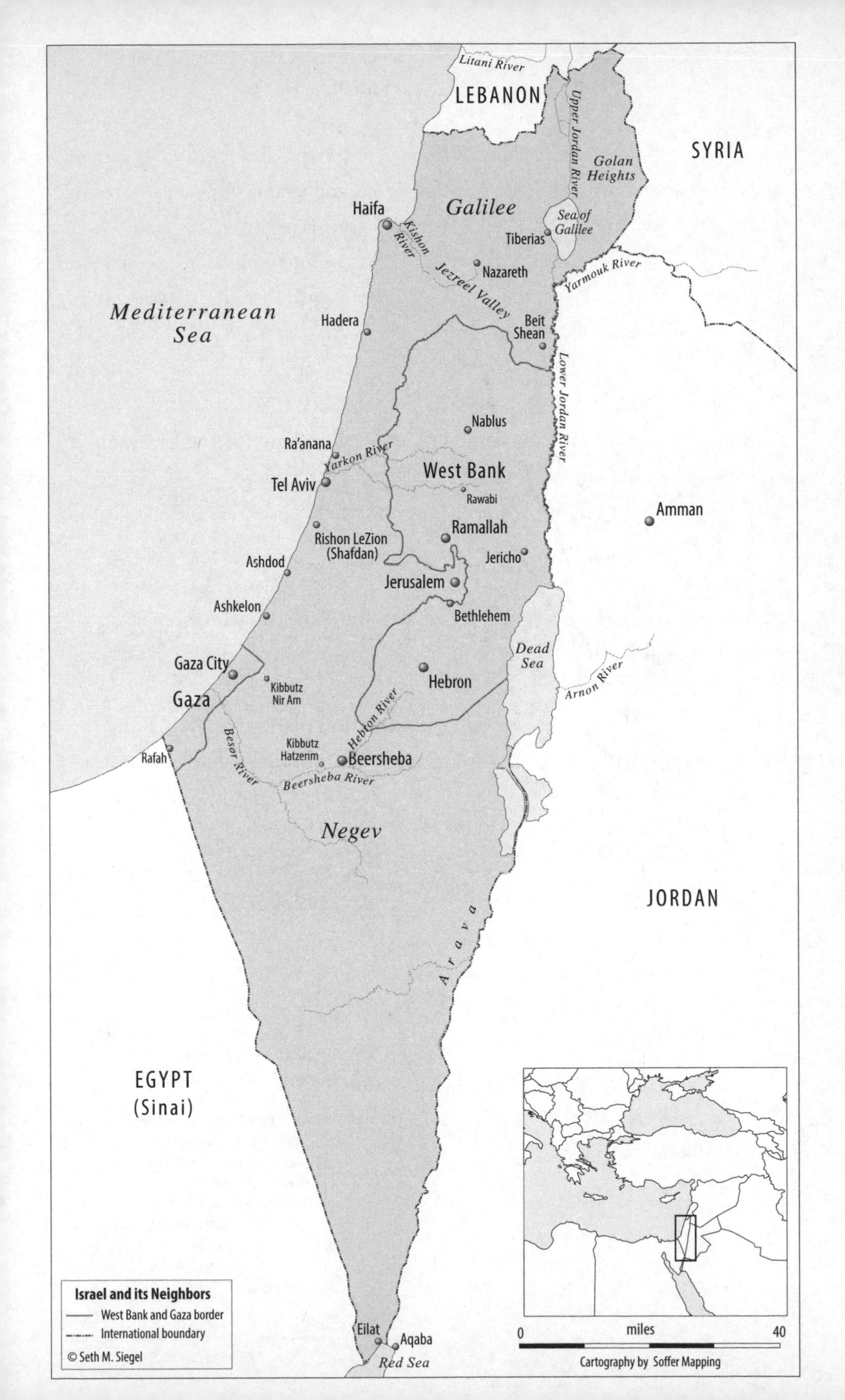

Litani River
LEBANON
SYRIA
Upper Jordan River
Golan Heights
Galilee
Haifa
Kishon River
Sea of Galilee
Tiberias
Nazareth
Jezreel Valley
Yarmouk River
Mediterranean Sea
Hadera
Beit Shean
Lower Jordan River
Nablus
Ra'anana
Yarkon River
West Bank
Tel Aviv
Rawabi
Amman
Rishon LeZion (Shafdan)
Ramallah
Jericho
Ashdod
Jerusalem
Ashkelon
Bethlehem
Dead Sea
Gaza City
Hebron
Kibbutz Nir Am
Gaza
Arnon River
Hebron River
Besor River
Kibbutz Hatzerim
Beersheba
Rafah
Beersheba River
Negev
JORDAN
Arava
EGYPT (Sinai)
Israel and its Neighbors
West Bank and Gaza border
International boundary
© Seth M. Siegel
Eilat
Aqaba
Red Sea
0
miles
40
Cartography by Soffer Mapping

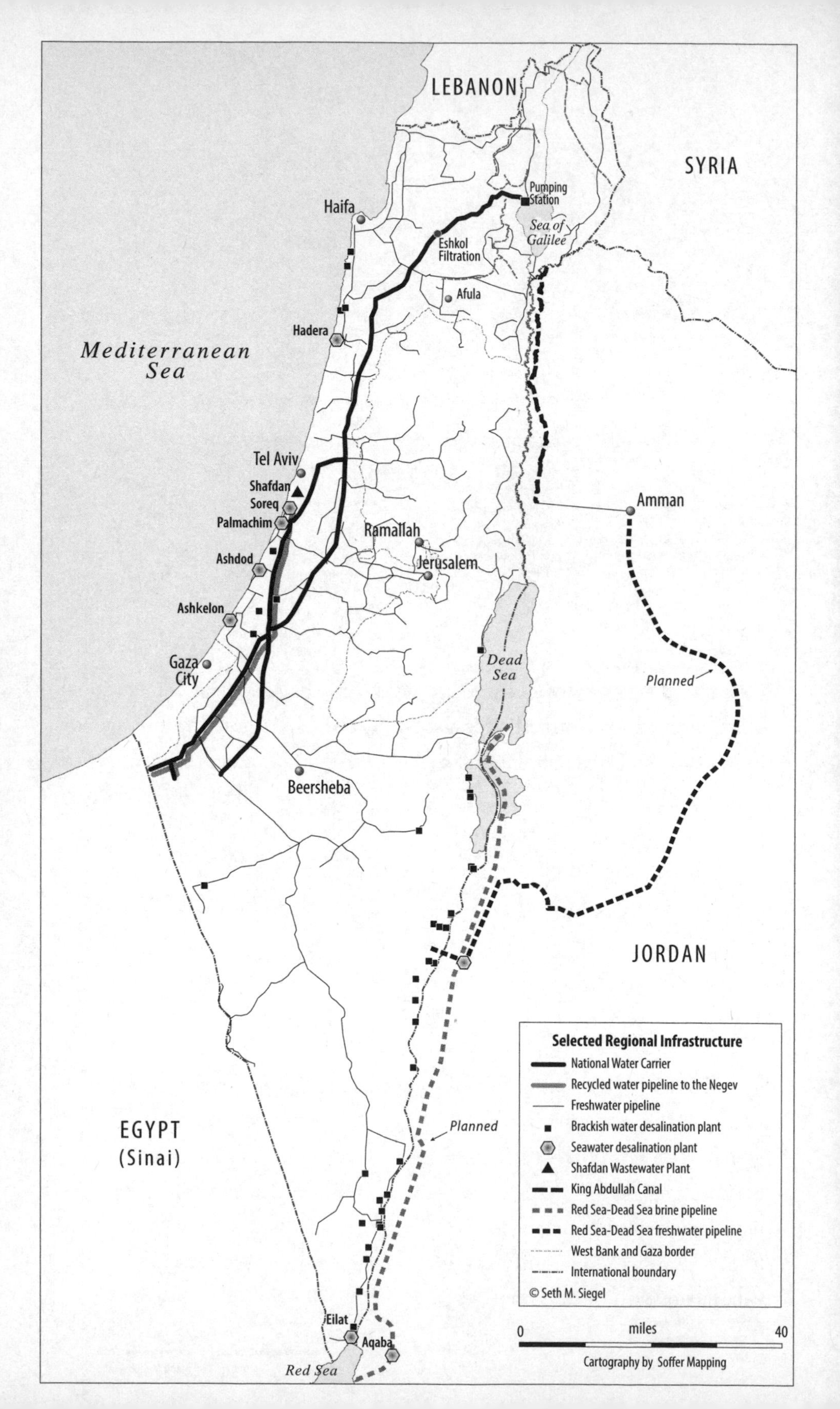
LEBANON
SYRIA
Pumping Station
Haifa
Sea of Galilee
Eshkol Filtration
Afula
Hadera
Mediterranean Sea
Tel Aviv
Shafdan
Soreq
Palmachim
Ramallah
Ashdod
Jerusalem
Amman
Ashkelon
Gaza City
Dead Sea
Planned
Beersheba
JORDAN
Selected Regional Infrastructure
National Water Carrier
Recycled water pipeline to the Negev
Freshwater pipeline
Brackish water desalination plant
Seawater desalination plant
Shafdan Wastewater Plant
King Abdullah Canal
Red Sea-Dead Sea brine pipeline
Red Sea-Dead Sea freshwater pipeline
West Bank and Gaza border
International boundary
© Seth M. Siegel
EGYPT (Sinai)
Planned
0
miles
40
Eilat
Aqaba
Red Sea
Cartography by Soffer Mapping

Introduction

A GLOBAL WATER CRISIS LOOMS

You ain't gonna miss your water until your well runs dry.
—Bob Marley

DESPITE ITS NAME, there are no covert operations at the National Intelligence Council. It is a sober, cautious US government agency more akin to a university faculty club or a think tank than the spy agency its name suggests. The council issues reports—some of which are top secret—that integrate information from other intelligence agencies to help government officials and other policy makers take a long view of coming problems.[1] So, it was odd that this conservative organization would issue a top secret, and then partially declassified, report with the provocative conclusion that the world is entering into a prolonged water crisis.[2]

The first parts of the crisis are already being felt. It is no longer a surprise to hear of a drought here, an overpumped aquifer there, or some social unrest in a country no one thinks about too often. But if the intelligence report is correct, the problem will soon begin to accelerate. It is less a matter of "if" than "when." In less than a decade, the report predicts, countries important to the US and to global security will be at risk of "state failure." The only questions in the report are how severe the disruptions will be and how quickly they will be felt.

Water shortages may not occur everywhere, but hardly anyone will remain unaffected for long. Twenty percent of the world population—about 1.5 billion people—will be the first victims of this world water crisis, and already six hundred million of them have begun to experience water shortages.[3] Ultimately, sixty percent of Earth's land surface will be transformed. To begin with, depleted water supplies will pose a risk to both US and global food markets, which will result in higher food prices around the world.[4]

Because extracting and generating energy is so water intensive, the intelligence report predicts, "water problems will hinder" its production.[5] This has already begun to happen in Brazil, South America's economic engine.[6] The report continues: "The ability of key countries to produce food and generate energy" will transform the world as we know it, "posing a risk to global food markets and hobbling economic growth."[7] As less energy is available, economic growth will slow. Higher food prices along with slower economic growth are a proven formula for social unrest.[8]

The water crisis isn't a "developing world" problem reserved for international aid organizations operating in faraway locales. Major trading partners of the US and world-economic powers like China and India are already experiencing water shortages that could soon have a major impact on their economies and political stability. And the water future in the United States—most immediately in the western states—is also at a tipping point. Water scarcities are morphing into full-blown water crises that, directly or indirectly, will affect nearly everyone in the US, whether in where they live, how much they have to pay for food, or how they earn their livelihood.

The San Joaquin Valley in California is the epicenter of high-end agriculture, with more grapes, oranges, peaches, vegetables, almonds, and pistachios produced there than anywhere else in the US. But already parts of the Valley are running out of water and the Valley as a whole is facing growing shortages.[9] The bounteous supply of California produce cannot be assured. Already, prices for these products have risen and water restrictions of increasing severity have been imposed on the once limitless California lifestyle.

California is not the only state in immediate danger. Since the end of World War II, the massive natural underground reservoir called the High Plains Aquifer has been a key driver of agriculture in eight Great Plains states. Basic crops grown there like wheat, corn, soy, and barley provide fodder for America's cows and chickens, and grains for food production. These crops are also a major US export industry. The aquifer that provides water for these crops has been so severely and consistently overdrawn that parts of it have already gone dry.[10]

Although the water in the High Plains Aquifer is a renewable resource, it took thousands of years of rain and snow to fill a large part of what has been depleted just since the 1950s when the overpumping began.[11] Worse, rather than slowing the depletion, in just the first years of this century, the High Plains Aquifer was further reduced by an amount equal to about a third of all of the water withdrawals made in the twentieth century.[12] The financial well-being and quality of life of millions of Americans will be affected, and not just of those farmers in Colorado, Nebraska, Kansas, Texas, and the other states where the loss of water is accelerating.

The water level in Lake Mead may soon be too low to permit further pumping, affecting the production of clean hydroelectric power for states in the Southwest.[13] As with California, many communities in Arizona and Nevada have already imposed water-use restrictions as a growing population has overwhelmed the available water supply, even as local water supplies continue to be overtaxed.[14]

It isn't just drought that threatens the US water future. Pollution, too, is limiting available resources. To cite one example, Florida's largest source of freshwater—the Manatee Springs and Aquifer—has been tainted by agricultural runoff and will require expensive treatment if it is to remain safe for drinking.[15]

Water and infrastructure crises are almost always avoidable—and elements of this crisis, too, may be held at bay with focused action by government, business, and civic leaders. Some countries will still enjoy a continuous supply of water, even if they will suffer the consequences of the world around them failing to do so. But there is no question that many nations will miss the warning, and not just the usual list of developing countries with resource and infrastructure problems. Water problems are a proxy for bad governance, and there is a lot of bad governance.

There are several macro trends—five highlighted here—that have been the main drivers of this imminent water crisis, many of which are a long time in the making. There is no sign that any of these trends are about to stop or even slow.

Population. World population keeps growing. Much has been made of falling birthrates in many countries, but that doesn't address the much longer life expectancy in most places, even as compared to just a few decades ago. The world's population is now over seven billion and isn't expected to level off until 2050 when it will reach 9.5 billion people.[16] No matter how little those extra 2.5 billion people eat or bathe, it is certain that finding, cleaning, and supplying that additional water for even basic needs will be a challenge.[17]

Rising Middle Class. The world population isn't just growing; it is growing wealthier. Hundreds of millions of people recently living in poverty have already risen out of it and into the middle class, a trend that will largely continue. There were 1.4 billion middle-class people in the world in 2000. That number grew to more than 1.8 billion in 2009. By 2020, the world's middle-class population is projected to rise to about 3.25 billion people.[18] This is good news for humanity, but bad news for the global water supply.

The daily showers, backyard pools, and green lawns these more affluent people will enjoy will be a further stress, but the much bigger withdrawal from our water supply comes from the eating habits associated with a middle-class lifestyle. People living in extreme poverty tend to have vegetable- and grain-based diets; middle-class people overwhelmingly have protein-based diets. Raising a pound of beef uses *seventeen* times more water than it does to grow a pound of corn.[19]

But being middle class isn't just about food. The energy needed for operating cars, air conditioners, computers, and other appliances that are now a normal part of middle-class life utilizes almost unimaginable amounts of water. Several gallons of freshwater are needed to produce each gallon of oil, whether produced at home or abroad. Natural gas and fracking require millions of gallons per site. With the US now a major energy producer, billions of gallons of water are consumed for it just in the US *each day*.[20]

Climate Change. Climate change causes surface temperatures of lakes and rivers to rise, which in turn causes faster evaporation.[21] Higher temperatures require more water just to get the same amount of irrigation to crops. Rain patterns are also changing: The interval between rainfalls has been growing as has the intensity when it does fall. The longer intervals without rain have led to a hardening of surface soil. When the rain does come, much of it runs to sewers and rivers, or sits on the surface before evaporating, but either way it is lost because the rainwater can't permeate the ground.[22]

Tainted Water. Pollution is also reducing the amount of available water. Growing food for so many people and feed for so many animals requires ever larger amounts of fertilizer and pesticides, a portion of which is carried by irrigation or rain into aquifers, lakes, and rivers. Energy-extraction techniques like fracking not only require the use of large quantities of water, but the chemical additives used in the fracking process are alleged to be polluting nearby drinking water reserves. Whether that claim proves to be correct or not, it is certain that all around the world, chemicals are seeping into water supplies. Some of these industrial compounds are carcinogens.[23] Regardless of how the water becomes tainted, undoing the damage to compromised aquifers and lakes is expensive and, at times, impossible. When water supplies become polluted, they are lost, sometimes permanently.

Leaks. Finally, a staggering amount of municipal water is lost every day in cities all over the world from leaks, open hydrants, theft, and neglect. London loses about thirty percent of its water and Chicago about a quarter.[24] Some major Middle Eastern and Asian cities can lose up to sixty percent of their system's water every year to faulty infrastructure; losses of fifty percent are not uncommon.[25] New York City has reduced its water losses due to leaks, but still loses billions of gallons, with one massive hard-to-repair leak losing thirty-five million gallons of water *every day*.[26] These losses may be invisible, but they are enormous.

Each of these challenges (population growth, rising affluence, climate change, pollution of water sources, leaky infrastructure, and others) can be overcome. Focus, will, creativity, trained personnel, and money are required. While every country should begin addressing these challenges, it is a near certainty that not every country will. But these problems can be addressed and even solved.

Rising demand and limited quantities don't have to restrict economic growth or lead to political instability. The lack of a natural supply of water, or declining rainfall, need not determine a country's destiny. If handled wisely, these limits can propel a country and create new opportunities.

A Model for a World in Crisis

Sixty percent of Israel is desert, and the rest is semiarid. Since its founding in 1948, the country's population has grown more than *tenfold*,[27] one of the world's fastest growth rates in the post–World War II era. Israel started poor, but now has one of the world's most rapidly growing economies.[28] Middle-class life is the norm in Israel. Its annual rainfall—not generous to begin with—has dropped by more than half.[29] Yet, despite its challenging climate and unforgiving landscape,[30] Israel not only doesn't have a water crisis, it has a water surplus. It even exports water to some of its neighbors.[31]

Let There Be Water is the account of how a small country developed its sophisticated approach to water, starting long before its independence. Water planning and technological solutions have been central to the country at every stage of its development. Even before it was a water powerhouse, Israel used its water know-how to help build relations around the world.

There are other countries that take water issues seriously and plan far ahead, especially Australia and Singapore. In the US, a few states like Nevada and Arizona have been planning for scarcity for a long time, even if both are constantly playing catch-up to the demand and the threat before them.

Of course, not everything Israel has done in terms of its water supply is relevant everywhere or to everyone. Countries with large landmasses are different in scale or topography from little Israel. Some countries have no desert or they have long rainy seasons or an abundance of lakes and rivers. Some countries' economies can't bear all the infrastructure spending that Israel incurs. Even so, some part of what Israel has done can help to transform the water governance of every country. If nothing else, Israel's focus on water and its priority in the national consciousness can be an inspiration to leaders and engaged citizens everywhere regardless of geography or affluence.

It would have been wiser for the world to have begun planning for water shortages and conserving water decades ago. But it isn't too late to begin now.

Here's how Israel did it.

PART I

THE CREATION OF A WATER-FOCUSED NATION

One

A WATER-RESPECTING CULTURE

Rain, rain, go away,
Come again some other day!
—American nursery rhyme

Rain, rain, from the skies
All day long, drops of water
Drip drop drip drop
Clap your hands!
—Israeli nursery rhyme

Aya Mironi, now in her thirties, remembers bath time when she was a little girl. As soon as she was toweled off and in her pajamas, her mother would return to the bathtub with a plastic bucket, and fill it with the water still in the tub. Her mother would carry the bucket outside to their small yard and water the flowers and other plants with the still soapy water. She would return to the bathroom, refill the bucket, and repeat the procedure several more times.

If you didn't know that this was taking place in an upper-middle-class community in Israel, you might have thought it was in a poor village in a developing country. Despite the free-flowing water in the home, though, Aya's mother treated their water as a precious asset not to be wasted. Over time, with one act of maternal water conservation after another, Aya and her two siblings absorbed the lesson that every drop counts. Once ingrained, it is a hard belief to unlearn.

Aya can also recall regular reminders in school to be mindful of water. There were posters in every classroom exhorting the children to "not waste a drop." She learned, as all Israeli children do, the sing-song Israeli nursery rhyme at the start of this chapter.[1] It is difficult to imagine an American child being coached to clap hands in delight because of a rainy day. In the American nursery rhyme, of course, the rain is shooed away to "come another day."

The wisdom of water conservation isn't limited to nursery songs. Rather it is part of an integrated curriculum that, like Aya's mother, tries to inculcate in schoolchildren both the idea that saving water is everyone's responsibility while at the same time giving them practical tools for doing so. Aya's mother may have been diligent in saving water, but the school program also trains children to teach those best practices to their parents. As part of hygiene classes, Israeli schoolchildren are taught to shower and brush their teeth, as students are in such classes everywhere. In Israel, there is an added feature: Students are taught how to minimize the use of water.[2] Saving water is everyone's job—but so is the educational process that leads to it.

The people of Israel aren't single-minded, water-saving zealots, but there is a general consciousness about the need to respect water and to not take it for granted. This water-conscious culture comes, in part, from the surroundings, with most of Israel being made up of desert and the rest semiarid land. Droughts are not uncommon. Even so, the physical environment alone doesn't adequately explain the heightened national consciousness regarding water and its preciousness.

Although most of the Jews in Israel today are not strict in their religious observance, culture and tradition are enduring phenomena.[3] The religious culture that carried the Jewish people for two thousand years from exile to national rebirth is filled with reverence for water in the form of rain and dew.

The prayers of Jews through those millennia and to this day include a prayer for rain at certain times of the year. This prayer is recited by many Jews three times each day in both the Diaspora and the Land of Israel. It doesn't ask for rain for the community where the prayer is uttered, but rather custom calls for it to fall in the Land of Israel. No matter where Jews are, in wet places or dry ones, their prayers have been recited for two thousand years facing toward Jerusalem—and with the meteorological well-being of the Holy Land in mind. As with Aya and her siblings, over time, this concern became ingrained and part of the Jewish communal worldview.

Separate from the prayer book, the Hebrew Bible also provides guidance on how to think about water. In one of the Bible's most famous scenes, in the midst of the wanderings of the Children of Israel, Moses strikes a rock in pursuit of freshwater to drink, and a "copious" amount flows.[4] This episode suggests a subtle division of labor: God provides nourishment for the Israelites with daily portions of *manna*, but it falls to Moses—even with Divine guidance—to provide water. The story is also a reminder that water may be found in many unlikely places and can sometimes be extracted with unorthodox techniques.

Each year, shortly before Rosh Hashanah, the Jewish New Year, Moses' blessings and curses from the Book of Deuteronomy are recited in every synagogue around the world. Rain "in its season" is one such blessing.[5] Perhaps the most famous of all Jewish prayers, the *Shema,* draws from Deuteronomy and states that a penalty for failing to observe God's commandments is that rain shall not fall and that lack of rain shall cause the violator to "perish."[6]

These water-focused scriptural episodes aren't outliers. Linguistically, the Hebrew Bible is a moisture-suffused document. The word *dew* is mentioned 35 times, the word *flood* appears 61 times, and the word *cloud* shows up 130 times. The word *water* itself is found 600 times in the Hebrew Bible.[7]

"Rain" is not only mentioned nearly one hundred times in the Jewish holy book, but there are even specific Hebrew words—still in use in modern Hebrew—for the first and last rainfalls of the year. If Eskimos have multiple words for snow because of its constant presence, Jews in the Holy Land would seem to have several words for rain because of its scarcity.

As the Zionist settlers were overwhelmingly secular, they may not have been dipping into their prayer books or Bibles with any regularity. But they arrived—from rainy lands like Russia and Poland and river-fed ones like Egypt and modern Iraq—with a familiarity with the Bible and Jewish tradition. From that, they had an inbred awareness of water from the enduring Jewish tradition around them that was tied to their new lives in the Land of Israel.

Water Engineers as Heroes

Theodor Herzl was a Viennese lawyer, journalist, and writer who—unlike many of the Zionist pioneers—knew little about Jewish tradition or custom. He had a quasi-spiritual Jewish awakening when he saw a spasm of broad-based anti-Semitism in genteel Paris in 1894. From this experience, the visionary Herzl came to conclude that Jewish life was doomed in Europe as Jews would fall prey to assimilation or persecution or both. He devoted the rest of his short life to the creation of the modern political Zionist movement.[8]

While building political support for a Jewish home, Herzl also wrote essays, plays, and books, all making the case for Zionism. The two most significant were a political tract, *The Jewish State*, in 1896, and a utopian novel in the style of the then best-seller Edward Bellamy's *Looking Backward*. Herzl called his 1902 novel *Altneuland*, or *Old New Land*.

As the Zionist movement had no religious works at its center, for many, Herzl's speeches, writing, and diaries assumed that place. Granted a secular holiness, Herzl's works were widely translated and any literate Zionist would have read, at a minimum, these two writings. When Herzl died at the age of forty-four in 1904, his insights were treated as guidance and inspiration from the grave. Decades later, Israeli leaders would still quote from Herzl and these books.[9]

In November 1898, the politically skillful Herzl arranged a meeting with the last German emperor, Kaiser Wilhelm II, to encourage his help in creating a Jewish state in the Land of Israel. The kaiser gave Herzl reason to think he would be an ardent supporter, praising the work of the Zionist pioneers. He told Herzl that above all else "water and shade [trees]" will restore the land to its ancient glory.[10] When Herzl's futuristic *Altneuland* was published four years later, one of the novel's lead characters says of Jewish settlement in Palestine, "This country needs nothing but water and shade to have a very great future."[11]

Late in *Altneuland*, one of Herzl's protagonists predicts that the water engineers of his imaginary Jewish homeland will be its heroes.[12] Herzl fantasizes about the country's water future. Although Palestine then was a place with meager water resources or cultivable soil, he describes its watery destiny and good fortune: "[E]very drop of water that fell from the heavens was exploited for the public good. Milk and honey once more flowed in the ancient home of the Jews. Palestine was again the Promised Land."[13] Utopian novels do set the bar high, and Herzl held the Zionist project, especially in regards to water, to that standard. So, too, did his political heirs.

Beyond books and exhortations, water entered the collective consciousness of the Zionist pioneers in other ways, as well. In one of the most enduring songs from the pre-State Zionist community, the pioneers often danced the *hora* circle dance to a water-themed song—as do many today, even far from Israel. The song *"Mayim Mayim"* [Water Water] is likely familiar to anyone who has ever been to a Bar or Bat Mitzvah party or a Jewish wedding. Although the lyrics come from the Book of Isaiah ("With joy, you shall draw water from the springs of Salvation"),[14] it was set to music and choreographed to celebrate the discovery of water at a collective farm in 1937 after years of drilling for water there had yielded only dry holes.

Other songs and folk dances were also composed and choreographed to celebrate water milestones.[15] While dancing the *hora* in the US may be reserved for Jewish celebrations, in Israel, until recently, folk dancing was an everyday form of socializing and exercise. Dancing to *"Mayim Mayim"* and these other water songs was a nearly universal cultural experience whether in the city or on the farm.

Water has also been used as a theme among leading Israeli writers, explicitly or metaphorically. In A. B. Yehoshua's novella *Early in the Summer 1970*, water is a motif running throughout the work. Dryness is synonymous with failed communication; the desert stands for barrenness and death.[16] Likewise, *My Michael*, Amos Oz's 1968 novel about life in 1950s Jerusalem, utilizes rain for symbolic impact. Rain and intimacy between the characters run in parallel while anticipation of rain is also used to literary effect.[17] More recently, Israeli novelist Assaf Gavron's dystopic, futuristic novel *Hydromania*, about life in Israel in 2067, uses water and rain as key plot devices to describe what happens when the people lose control of this essential natural resource.[18]

Israel has even honored water on its currency and stamps. The now-out-of-circulation five-shekel note (worth a little more than a US dollar today) featured Israeli Prime Minister Levi Eshkol. On the bill's back, Israel's National Water Carrier, a project in which he played a key role, was honored. Similarly, many Israeli postage stamps celebrate water themes ranging from technological innovations in water usage to milestones in modern water infrastructure to ancient water systems in the Land of Israel.

The Water Belongs to All of the People

No decision made by the Zionist pioneers and the young State of Israel has had a greater impact on Israel's water culture than the decision to make water the common property of all. Unlike in the US, where water is a personal property right, in Israel all water ownership and usage is controlled by the government acting in the interest of the people as a whole. Available water is then allocated according to what is seen as the best use.

The control of the nation's water was codified in a series of laws that confirmed the centralized water philosophy of Israel. These water laws also came to play an essential part in Israel's water-conservation success.

In the mid-1950s, three laws were passed by the Knesset, Israel's Parliament, which set the stage for the transformative Water Law of 1959. The first, passed in 1955, prohibited any drilling for water anywhere in the country, even by an owner on his private land, without first obtaining a license to do so.[19] Private property rights yielded to government control.

The second of the water laws, also passed in 1955, prohibited any distribution of water, unless that supply was done through a meter.[20] This law also required that all utilities install separate meters to measure the amount of water provided to each home or business.[21] While this granular collection of data put Israel decades ahead of the information technology boom (and the metering infrastructure would prove to be of immeasurable value in later years), it again established an intrusive government role in the water-consumption patterns of its citizens.

In 1957, a third water law was passed by the Knesset. With control of underground water spoken for in the 1955 water-drilling legislation, this new law addressed surface water, broadly construed. Not only did this place the water found in rivers and streams under government control, but it also took charge of rainwater. It even took ownership of the sewage flowing out of Israelis' homes.[22] The law prohibited diversion of any of these forms of water without first receiving a government permit.[23] It also compelled farmers to obtain a license before herding their own grazing animals on their own property if the animals would cross a waterway in the process.[24] Individual interests, yet again, were subordinated to government control.

This evolving centralized ownership reached its logical culmination with the Water Law of 1959. The legislation vested in the government "widespread power to control and restrict the activities of individual water users in order to further and protect the public interest."[25] All water resources became public property subject to control of the State.[26] Land ownership would confer no rights to water resources on, under, or adjacent to the owner's land.[27] Henceforth, individual or private use would only be permitted if in accordance with the law.[28] The Water Law even stated an expectation that all citizens would use the water they receive "efficiently and sparingly."[29]

While popular acquiescence in this state control can be understood in the early years of the country, when the government had a decidedly socialist tilt, it might be expected that the Water Law would have been amended or repealed as the country abandoned its socialist origins. However, ownership of water continues to remain exclusively in the hands of "the people"—and therefore, of the government. Even after several rounds of privatization of government-owned industries and assets, there has been no call for water resources to become a free-market commodity. Israel today has a dynamic capitalist economy, but with a state-controlled, centrally planned approach to its water.

Shimon Tal, Israel's water commissioner from 2000 to 2006, provides a vivid illustration of how completely water is under the power of the state in Israel. "Of course, the government controls all of the water in the Sea of Galilee [Israel's largest freshwater lake] and of course, it controls all of the aquifers," he says. "But if you put a bucket on the roof of your house at the start of the rainy season, you own the house and you own the bucket, but the rain in that bucket is the property—at least in theory—of the government. Without a license to collect that rainwater, you are technically in violation of the Water Law. Once the raindrop hits the ground, or the bucket, it is owned by the public."[30]

Even compared to other countries with public ownership of water, Israel has taken a more absolutist approach than most. In France, for example, a landowner doesn't have unfettered right to use all of the water under his land to the detriment of others. But the 1964 French water law says he can use that water freely provided he doesn't deprive his community reasonable access.[31] Further, the French Civil Code explicitly gives ownership of rain to the owner of the land where the drop falls.[32]

A visitor to Israel might assume that such a controlling, restrictive law and policy was an unpopular one, especially in a country that has seen the near collapse of its socialist political parties and a general repudiation of socialist economics. But it is the opposite. Israelis widely believe that the collective approach in this instance is the secret to the nation's success in water conservation.

Professor Arnon Soffer is a political geographer and the founder of the Department of Geography at the University of Haifa. He studies water systems all over the world. He is also, philosophically, a supporter of free markets and not fond of government intrusion. Still, he says, Israel is "a Western country and we embrace the idea of individualism here. But there are some areas where the kibbutz [collective] approach makes the most sense. With water, collective ownership is one of the reasons why we are able to be a villa in a surrounding jungle."[33]

Israelis have accepted a trade-off. They have surrendered private ownership and the benefits of a market economy in water for a system that offers universal access to high-quality water. The public gives the government the power to manage, regulate, price, and allocate water in its name with the belief that the common good will be the greatest beneficiary.

Israel's water system may be the most successful example of socialism in practice anywhere in the world today.

Two

THE NATIONAL WATER CARRIER

Water to the country is like blood to a human being.
—Prime Minister Levi Eshkol

THERE WAS NO CRISIS that so tested the Zionist cause as the British White Paper of May 1939, a government decree issued to choke off Jewish immigration to Palestine, the area that today comprises Israel, the West Bank, and Gaza.[1] While the British largely achieved their goal, the White Paper had an unintended effect: It led to new thinking by the Zionists about how to manage the nation's water for the broadest benefit, culminating almost exactly twenty-five years later in June 1964 in Israel's National Water Carrier.

The National Water Carrier was a feat of imagination and daring, requiring engineering innovation and a variety of financing vehicles, including one that led to riots and deep divisions that took years to heal. But, the planning and construction of the national water infrastructure also served to unify the nation while transforming the country.

Beginning in 1936, the British authorities faced three years of Arab riots in Palestine, a territory they had ruled since the end of World War I. The ostensible reason for the disturbances and bloodshed was the rising Jewish immigration, but if Jews were the initial target of the Arab rioters, the British police and military soon also became their focus. By 1939, the intermittent riots seemed to have come to an end, but the British Foreign Office in London was worried about a revival of these uprisings.

Concerned that war might soon break out in Europe, British officials did not want to have to tie down troops to keep the calm in Palestine. They were also keeping an eye on other potentially restive Muslim communities in their far-flung colonies and wanted to be sure that none might use disturbances in Palestine as a pretext for their own anti-British, pro-independence rebellions, which would also distract from the war effort. Making sure there would be no repeat of the 1936–1939 Palestinian Arab revolt became a key foreign policy interest of the British Foreign Office.[2]

These imperial fears dovetailed with a concern first expressed by British economists in the late 1920s that Jewish immigration to Palestine was not sustainable and would soon overwhelm available water resources for agriculture and other uses. The economists were of the belief that the entire geographic area of Palestine could hold no more than two million people. With natural increase, Palestine's 834,000 total inhabitants in 1939 would reach that threshold in a generation or so, but it would be reached even sooner if open immigration would be adding to the ranks of the 150,000 Jews already living there. In looking at the competing interests of the Zionist movement seeking to promote immigration versus the fragile ecology and limited water resources of a region the British hoped to long rule, Prime Minister Neville Chamberlain's government came up with a solution in the 1939 White Paper that the British hoped would also placate the local Arab population.

Under the terms of the decree, Jewish immigration to Palestine would be limited to 75,000 people over five years, a mere 15,000 per year.[3] With Jewish emigration and natural deaths, the Jewish population would likely be at or about its then current level five years hence. The Zionist effort at creating a Jewish state would be killed in its cradle.

While the White Paper has been widely analyzed from a political perspective as well as on its tragic consequences for European Jews seeking refuge from the Nazis in World War II, it is also a valuable starting point in understanding the modern State of Israel's approach to water. Most immediately, the Zionist leaders were desperate to demonstrate that the British economists' calculations on the amount of available water were wrong. More broadly, and for their own purposes, as well, the Jewish leadership needed to be certain that Palestine could hold millions more than the two million assumed as the maximum population by the British.

Beginning with the issuance of the White Paper, through the war years, and in the postwar period until the State of Israel was declared in May 1948, the Zionist leadership developed a series of plans demonstrating that the Land of Israel had great water potential, but which could be realized only with significant changes in the way water was then found and used. These plans were to no effect in changing British views or in increasing the number of Jewish refugees permitted to immigrate. But this new thinking and the plans that followed from it, set the philosophical and practical groundwork for Israeli water management that has permitted Israel to consistently, if not always perfectly, be ahead of the water needs of the country to this day.[4]

The geographic area of Palestine today is home to more than 12 million people, with about eight million in Israel and another four million or so divided between the West Bank and Gaza. In addition, Israel provides large amounts of water from its own supplies to both the Palestinians and the Kingdom of Jordan, and even exports billions of dollars each year of peppers, tomatoes, melons, and other water-intensive produce. That the British economists were wrong goes without saying.

Simcha Blass, the Water Man

If the world were a fairer place, Simcha Blass's name would be widely known in Israel and around the world. Town squares would be named for him and academic conferences would hold retrospective analyses on the role he played in changing Israel's water destiny. Now mostly forgotten to history, Blass was the central character in leading the thinking and planning about Israel's water, and later in transforming agriculture around the world.

By the early 1930s, Blass, a recent immigrant from Poland, was already on his way to being known as a water engineer of uncommon insight, intuition, and skill. Even so, the life of a water expert in the *Yishuv*, as the Jewish community in Palestine was known, was still fairly basic: drilling for water, pumping it to the surface, and conveying it short distances through pipes of narrow diameter. With or without the insights of the British economists, it was clear that, without changes, the water supply on hand would not be enough to accommodate the potential Jewish immigration, especially after February 1933 when the rise of Adolf Hitler and his Nazi regime created an ever-more urgent interest in European Jews immigrating to the Land of Israel.

The millions of people projected to arrive in Palestine—whether due to Zionist ideology or to the simple search for a safe port during a European storm—would need water for agriculture, industry, and for simple household use. The flow of water was as important as the flow of immigrants. One was tied to the other.

If Blass was the most significant water engineer in the *Yishuv*, he had an important partner in Levi Eshkol, who had held a variety of important positions in the pre-State Zionist political structure and who was a trusted senior aide to David Ben-Gurion, the political leader of the Jewish community in Palestine. Despite Eshkol's many responsibilities, no task animated him more than did water.[5] Eshkol, who would end up as Israel's third prime minister and head the country during the June 1967 Six-Day War, arguably has no greater legacy than creating the political and institutional framework for developing the country's major water infrastructure.

Beginning in the 1920s, the Zionist leadership created many organizations of the kind that would serve as pre-State institutions.[6] In the case of water, Eshkol teamed up with Blass and a few others in 1935 to plan the creation of a water company—called Mekorot at its founding two years later.[7] The company was responsible for water exploration and for assuring that water would be on hand when needed for the growing ranks of Jewish settlers and farmers in the British-controlled territory.

Even before Mekorot was formed, Eshkol asked Blass in 1935 to identify new water resources for the western Jezreel Valley, a Jewish farming district south of Nazareth and the Lower Galilee, which was experiencing rapid growth. After a series of successful well drillings led by Blass, water was soon found and was pumped to farms throughout the valley. Immigrant farmers in the Jezreel Valley were able to expand and new farms were soon added.[8]

As important as was the finding and transportation of the water, there was something far more consequential about the Jezreel Valley project: It was the first time that Blass had been called upon to create a plan to develop water resources for farms relatively far from the water's source. In the coming years and over far greater distances, Blass would develop grander water plans and execute projects which, cumulatively, would open ever greater parts of the country to productive use of the land and the production of more food for a soon-to-be growing nation.

A "Fantasy Plan" Changes How Water Is Managed

At the time the British White Paper was issued in May 1939, and the success of the Jezreel Valley water exploration notwithstanding, most of the *Yishuv*'s water for agriculture or household use came from shallow wells drilled in towns and farms along the Mediterranean coast. Water was mostly a balkanized, district-by-district affair with very little sharing or pooling of resources. As was then true throughout the region and most of the world, farms and towns would mostly use the water at hand with little pumped to where it might be put to better use.

In terms of volume of available water, the greatest existing water resources were in the far north of the country. There were a scattering of settlements and farms there, especially along the border with Lebanon and Syria, but it wasn't where the water was needed most.[9] The mass of population was clustered around Israel's new metropolis, Tel Aviv, in the center of the country's long coastline. The vast, open expanse of the Negev, largely unpopulated except for a few nomadic Bedouin tribes, was desert. But Ben-Gurion presciently believed the Negev offered the best hope for agriculture for the emerging state if water could be found for it.[10] In any case, neither the Tel Aviv area nor the Negev desert then had enough water to sustain the population growth Ben-Gurion had in mind.

Blass was asked to create a "fantasy water plan" that could be presented to the British in the hope that it might modify their thinking about expanding the number of Jewish immigrants. He immediately set to work. His idea was to develop a massive infrastructure project that would take water from the water-rich north and bring it to the water-limited center and water-impoverished south.

By July 1939, Blass had completed the first of what would be many drafts of his water plan, a design he would continue to revise for nearly twenty years, even long after the State of Israel was established and all restrictions on immigration lifted. His initial draft would evolve into the nation's water master plan, but all of the elements found decades later—including the National Water Carrier—were there from that first document. Everything thereafter would be commentary, elaboration, and execution.

Blass proposed a three-phase approach to national self-sufficiency in water. First, he believed there were large amounts of water below the surface of the Negev desert that could be found by deep drilling. In his plan, this water would be utilized almost immediately to establish up to thirty new farming settlements in the Negev. Second, he proposed pumping water out of the Yarkon River, north and east of Tel Aviv, and transporting it to the Negev, primarily for agricultural use. And then, at some undetermined time in the future, water would be brought from north to south via mostly underground infrastructure that would bisect the nation. That was to be the National Water Carrier.[11]

The "fantasy" element to Blass's plan was whether the British would be prepared to go beyond the borders of Palestine.[12] Only a short distance away, the Yarmouk River, in what was then Trans-Jordan, and the Litani River in Lebanon, uselessly drained large amounts of river water to the Jordan River and the Mediterranean Sea, respectively. With access to that unused water, the *Yishuv*—and millions of European Jews desperate to come—would have all of the water they could use.

Less than two months later, Germany invaded Poland and World War II began. Although the war made immigration far more difficult, there were still Jews eager and able to leave, if a visa could be found. Ben-Gurion continued to try to get the British to accept more refugees, and Blass's ever-evolving plan was a part of those requests.[13]

As Blass developed his ideas with ever greater nuance and detail, he tracked every water source inside or close to the boundaries of the Land of Israel and hypothesized a unitary national system with water flowing everywhere on demand. In a 1943 revision of his original plan, he detailed gathering the northern headwaters of the Jordan River and the Sea of Galilee, adding in the streams that flowed here and there, and also incorporating the ad hoc system of coastal wells. Using the diversion of the Colorado River as a model, an engineering feat which brought freshwater to Los Angeles, Blass created plans to transport these sources of water south, as needed, until the system would terminate at farms dotted throughout the then sparsely populated Negev.[14]

Further Blass drafts added features like trapping and collecting storm water; treating and reusing sewage for the health of the region's rivers and for the agricultural potential of such reuse; and more sophisticated drilling into the aquifers. He even offered a water-diversion plan—never executed—to build a canal from the Mediterranean Sea to the Dead Sea, utilizing the altitude drop to generate hydroelectricity.[15]

After Blass's several plans, the water genie was forever out of the bottle. Now, everyone in the *Yishuv* leadership knew that the Zionist project would move forward with an integrated, national system of water-resource management unlike anything seen before in the Middle East or in much of the world until that time.

The Zionists may not have had the political autonomy to do what they wanted with the British still their political masters. They may not have had the funds necessary to begin so grand a project. They may not even have known where the borders of their future state would lie. But there was no question that Blass's plans offered a way forward in building the water infrastructure needed for their modern state and for the absorption of millions of new immigrants.

The "Astonished" Walter Clay Lowdermilk's Bestseller

Simcha Blass wasn't the only one thinking about water plans for the Land of Israel.

An American soil scientist, Walter Clay Lowdermilk, was sent, in 1938, by the US Department of Agriculture, to make a comprehensive survey of the soil of Europe, North Africa, and Palestine. The point of the project was to see what could be learned from the soil of these old civilizations for application in American soil conservation efforts.[16] In February 1939, with war in Europe still more than half a year from erupting and the White Paper still a few months away, Lowdermilk arrived in the Land of Israel.

Lowdermilk was appalled by what he saw. Ancient terraces and topsoil had been largely eroded, washed to the Mediterranean Sea by millennia of neglect. But he was also "astonished"[17] by the soil reclamation efforts of the Zionists. Having just nearly concluded his fifteen-month tour visiting twenty-four countries, Lowdermilk described the agricultural restoration in the Land of Israel as being "the most remarkable" work he had seen in his long journey. He extended his stay to visit many of the three hundred farms, settlements, and outposts of the *Yishuv*. He drove more than twenty-three hundred miles within the Land of Israel and another thousand in Trans-Jordan.[18] The more he saw, the more he became enamored with the Zionist mission. Looking at inbound Arab migration to Palestine, and the rising Arab prosperity and falling Arab infant mortality, Lowdermilk saw Jewish settlement as a positive for both Arabs and Jews.[19]

Returning to the United States, Lowdermilk became excited by the opportunity to revitalize the Land of Israel for its own sake and as a template for agricultural and economic development in North Africa and throughout the Middle East. In 1944, as World War II was winding down, a major US publisher[20] released Lowdermilk's *Palestine, Land of Promise*. The book went through eleven printings and became a bestseller.[21] It received positive reviews, including one in *The New York Times*,[22] and a long, glowing one on the front page of the *New York Herald Tribune*'s weekend book section titled "The Miracle That Is Going on in Palestine: The Jews Restore Fertility Where the Desert Had Crept In."[23]

Lowdermilk's book made the case for a massive public works reclamation project in the Jordan River Valley that would marshal water supplies for irrigation, rebuild the topsoil, develop hydroelectric power, and reforest a land last heavily wooded about two thousand years earlier during the last Jewish Commonwealth in the Second Temple era. With all of that implemented, Lowdermilk believed that the Land of Israel would have adequate developable natural resources and could soon absorb four million Jewish refugees.[24]

More significantly for that moment, Lowdermilk rejected the prevailing White Paper doctrine that geographic Palestine had a specifically constrained population limit and took a swipe at the British: "The absorptive capacity of any country is a dynamic and expanding conception. It changes with the ability of the population to make maximum use of its land, and to put its economy on a scientific and productive basis."[25] Already in his first visit in 1939, Lowdermilk saw examples of sophisticated water technology at work by the Zionists and understood what it would mean.

Concluding his book, Lowdermilk showed great optimism about what the region could become: "If the forces of reclamation and progress Jewish settlers have introduced are permitted to continue, Palestine may well be the leaven that will transform the other lands of the Near East. Once the great undeveloped resources of those countries are properly exploited, twenty to thirty million people may live decent and prosperous lives where a few million now struggle for a bare existence. [The Jewish settlement in] Palestine can serve as the example, the demonstration, the lever, that will lift the entire Near East from its present desolate condition to a dignified place in a free world."[26]

Lowdermilk had a model in mind for how water could be utilized in Palestine with the Tennessee Valley Authority (TVA), President Franklin Roosevelt's Depression-era project to bring electricity and redirected water to a large swath of impoverished, rural America. Ben-Gurion had known about the TVA, and was impressed by its scope and audacity. As with Lowdermilk, Ben-Gurion wondered if the TVA could be replicated in the Land of Israel. An intermittent conversation led by Ben-Gurion and inspired by the TVA about a massive water project gained urgency with the restrictions of the White Paper. Lowdermilk was an advocate for adopting the ideas of the TVA, and even if his more-ambitious plan had points of difference with Blass's, it was a respected validation for its essential elements.[27]

If Lowdermilk had an influence on thinking about water in the Land of Israel, he may have had an even deeper impact on the embryonic thoughts about the Zionist effort among US policy elites. Lowdermilk's book was given to every member of Congress.[28] More remarkable, *Palestine, Land of Promise* was likely the last book President Roosevelt read; it was found open on his desk when he died.[29]

Not surprisingly, Lowdermilk was celebrated in the *Yishuv* and he ended his career, after the State of Israel was already formed, on the faculty of the Technion—Israel Institute of Technology.[30] Lowdermilk's Israel TVA plan furthered the conviction that Palestine could have huge water resources and confirmed the Zionist idea that with water, a large population could follow.

An Essential Wasteland for Ben-Gurion

From today's vantage point, the British, after World War II, were largely seen as exhausted, demoralized, and bankrupt, eager to evacuate their colonies and to bring the two-hundred-year-old British Empire to a close. That may have been true in some locales, but for British Foreign Secretary Ernest Bevin and the British security establishment, Palestine was not one of them.[31] Because of British interest in protecting the Eastern Mediterranean and safeguarding the Suez Canal for the safe passage of goods from India and oil from the Persian Gulf, Bevin was determined to stay.

Aside from the canal, the British had completed an Iraq-Mediterranean oil pipeline during the interwar years with Haifa serving as a geopolitical and strategic node. From Haifa, British oil tankers would load and make the short trip across the Mediterranean to England to literally fuel the revival of England's economy. After fifty years in Palestine, the British, or at least Bevin, were making plans to stay in Palestine for another fifty years.[32]

The Zionist leadership had other plans. To them, it was only a matter of time before economic and political pressures would force Great Britain to depart, at which point there would be a battle—military or political—over the borders of the new Jewish state.[33] While the *Yishuv* leaders would do all they could to secure the largest logical footprint, Ben-Gurion had a special interest in the Negev.[34] He was determined to do what was necessary to assure Jewish control over it whenever the day would come when the British packed up and left.[35] Before that would happen, the nascent United Nations would be assigned the task of drawing the borders of the Land of Israel.

For most observers, the Negev was inhospitable to human habitation, a wasteland. It was too hot in a pre-air-conditioned world for large numbers of people and too dry for agriculture. There seemed to be no water resources. But for Ben-Gurion, the Negev held several attractions. It would protect Israel from isolation by giving it a port on the Red Sea.

It would provide strategic depth against an Egyptian invasion via the Sinai Peninsula. And, once water issues were sorted out, it would provide a mostly unpopulated area for growth and land for farming.

Ben-Gurion was sure that without a toehold in the Negev, the United Nations would never give control over the Negev to the Zionists. He saw himself in a race against time to establish facts on the ground that would justify a recommendation by the UN commissioners to grant the desert territory to the new Jewish state. And this would put Phase One of Blass's plan—deep drilling for water in the Negev—to the test. But first, the Zionists would need to establish Jewish claims to the desert.

"The Champagne Pipeline"

On the night following Yom Kippur 1946, the Zionist leadership pulled off one of the most daring episodes of its cat-and-mouse struggle with the British over their continuing restrictions on immigration and settlement building. Water played a key part in this dramatic, nearly cinematic, event.[36]

Yom Kippur is unique in the Jewish calendar. For many, it is a day of fasting, prayer, and contemplation. For a few others, Yom Kippur 1946 was a day for making final preparations to challenge the British in a way never done before. At nightfall, when the holy day came to an end, eleven convoys set out for predetermined locations across the northern Negev.

Working under the cover of darkness, each of the teams built at least one structure and made sure to complete the roof of each building before sunrise. Under British law, Jews were forbidden to establish new farms and settlements in Palestine, but there was a loophole: an Ottoman law that was still in effect, predating the British conquest of Palestine, held that no structure with a roof could be demolished by the government unless it was a safety hazard.[37]

By that next morning, there were eleven new farms along the northern fringe of the Negev. Not one had been interrupted by British intervention whose army had likely let down its guard because of the Yom Kippur holiday. (Even better luck for the Zionists, Yom Kippur came to an end on a Saturday evening when British troops often spent the night drinking, followed by Sunday mornings sleeping it off.) The settlers achieved their initial goal of getting the farms established.

Despite this overnight success, all eleven farms were missing one essential ingredient: water. Each of the convoys came with a water truck, but that was only a stopgap measure. Without significant amounts of water, these farms would soon wither. Water trucks might be enough for daily life, food preparation, and sanitation. But any crops they hoped to plant would not survive long without water for irrigation.

Simcha Blass had been part of the planning team for the eleven settlements, helping to select their locations in places either most likely to have water underground or within a pipeline's distance from a source. Now, it would be up to him to see if these farms could endure. The first phase of his three-phase water plan called for drilling wells in the Negev—likely to significant depths—in search of local water supplies. He began drilling and at Nir Am, one of the eleven new farms, water was found.[38]

Blass had a problem, though: He needed hardware to move the water. World War II had created massive shortages of metal and machinery, with most industrial goods allocated to the war effort. In the Land of Israel, many of Blass's projects were hindered by a scarcity of pumps and pipes. In the aftermath of the war, shortages continued as there was a seemingly endless demand by the civilian sector in the US and the effort to rebuild all of war-ravaged Europe. Anticipating the need to pump water to these eleven farms, Blass had quietly made arrangements to purchase a large consignment of steel pipes from an unlikely source.

During the war, a special set of pipes had been laid to help put out fires caused by the Nazi blitz of London. With the war over, and the Nazi threat gone, this parallel London water system was superfluous. Blass quietly arranged to purchase all of those pipes. The expense was enormous, but high-quality pipes were hard to find. With his new trove of hardware, Blass was able to arrange for the desert farms to be linked up to Nir Am. As with his 1935 project in the Jezreel Valley, Blass had established a regional water system that would have long-lasting impact on the Zionist cause and also on the to-be-formed nation's approach to water.[39]

It was an episode filled with irony. The discarded British pipes first used to frustrate Hitler's effort to terrorize the people of London now served to undermine British efforts to stymie Jewish settlement construction. Because of how much the pipes cost, the Negev infrastructure was dubbed the "Champagne Pipeline."[40] For the *Yishuv* leaders, and certainly for Ben-Gurion, almost any cost would have been worthwhile if it solidified the Zionist hold on the Negev.

Paying for the National Water Carrier

Ben-Gurion was to get his way.

In 1947, the United Nations sent a committee of experts to Palestine to study how the land should be divided. With the Negev marginally settled by Jewish farmers and no superior claims by others,[41] the committee awarded the desert wasteland to the still-unnamed Jewish nation, making well more than half of the country's territory seemingly worthless and uninhabitable. The British also gave testimony to the UN delegates, and reiterated their belief that the territory could not provide for the many homeless, stateless Holocaust survivors then in refugee camps in Europe, two years after the war had come to an end.

The ever present Simcha Blass was brought in by the *Yishuv* leadership to counter the British view. He presented his three-phase plan, and how Phase One had successfully brought drilled Negev water at Nir Am to the eleven Negev farms. Apparently, his descriptions of the still entirely fanciful Phase Two (bringing Tel Aviv's Yarkon River water to the Negev) and Phase Three (his Robin Hood plan of taking water from the water-rich north and giving it to the water-poor south) persuaded the UN investigators. They left rejecting the British assumptions and accepting Blass's estimates that the Land of Israel could have water resources nearly triple the proven amount then on hand.[42]

The State of Israel was declared on May 14, 1948, and the armies of six Arab countries invaded the newly born state.[43] Water issues were shunted to the side with national security demanding everyone's time and attention. After a cessation of fighting and the signing of armistice agreements in the first half of 1949, a flood of Holocaust survivors from Europe and Jews from Arab countries now facing persecution in their homelands began arriving in Israel.[44]

On the day of the declaration of Israel's independence in 1948, the national population was 806,000.[45] In the ensuing three-and-a-half years, more than 685,000 immigrants arrived in the new country.[46] Likely no nation has ever absorbed such a large percentage of its base population in so short a time. To grow food for this nearly doubled population, and to provide employment for many of the newcomers, new farms were set up in every corner of the country. More than for household use, water for agriculture was desperately needed.[47]

Phases Two and Three of the Blass plan were still only detailed concepts, intended for later action. Before substantive steps could be taken, financing would need to be secured. The combined costs of having fought a multifront war with continuing security burdens and absorbing penniless Jewish refugees still streaming in from Europe and the Arab world had thrown the country deep into debt and necessitated the rationing of food. Even so, Ben-Gurion and Eshkol were eager to begin building the water infrastructure.

Although doing so sparked riots and talk of civil war, Ben-Gurion accepted a proposed reparations agreement with the postwar West German government, which agreed to pay reparations to Israel both for resettlement costs of displaced survivors and also for the billions of dollars of Jewish property that had been stolen or destroyed by the Nazis. Many inside Israel did not want to take what they saw as blood money. Despite the upheaval within Israel over accepting anything from the land of the former Nazi regime, again, Ben-Gurion had his way. In a close vote, Israel's Parliament ratified the agreement with Germany.[48] Funding to begin building water (and other) infrastructure was now in hand.

Winning Over the Movie Executive

Even with German reparations payments beginning to arrive in early 1953, the Israelis were still missing an essential ingredient for their north-south pipeline to the Negev: a guaranteed supply of water. While Blass was sure that there was an abundant quantity of water available, Israel's hostile neighbors would have something to say about Israel taking water that also bordered their states. Protocols had to be established confirming who could take what from the Jordan River and its tributaries.

After a series of military skirmishes primarily between Syria and Israel, but also involving Jordan and Lebanon, President Dwight Eisenhower decided to use the hostilities as an opportunity to get the US involved. While the distribution of water was a concern, the World War II Supreme Commander thought in broad geostrategic concepts. Eisenhower's greater interest was that the Soviet Union not exploit Arab-Israeli tensions as a way of shoehorning itself into the region.

Eisenhower hoped that in negotiating a resolution of a technical but vital issue like water, tension surrounding the Arab-Israeli dispute and the Palestinian refugee issue could be eased, even if not resolved.[49] Rather than pick a diplomat to head the effort at resolving the conflict, Eisenhower turned to Eric Johnston, the head of the Motion Picture Association of America and a leading Republican who had experience in international development, and made him his special ambassador to the region.[50]

Johnston arrived in October 1953 with a plan for dividing up the water of the Jordan River. As soon as it was presented to the Israelis, it was clear that any dream of transporting water from the north to the Negev would die with Johnston's plan. Among much else, the American proposal had two particularly troubling features. First, Johnston wanted to allocate a far smaller amount of Jordan River water to Israel than Israel believed it deserved and certainly less than it would need to enable the Negev to bloom with new farms and fields. Second, Johnston arrived with the point of view—shared by the Arabs—that all of the Jordan River water would have to stay in the Jordan River basin for that region's development. In other words, even if more water could be found, Israel wouldn't be permitted to send it to the Negev.[51]

Blass was enlisted to serve as tour guide and tutor of Johnston.[52] As time went by, Johnston reversed his position on both of these principles, either of which would have doomed the National Water Carrier. First, he came to see the wisdom of using all available water resources "without undue waste, and that the volume of crops that can be grown in the region should be the paramount criterion of desirability."[53] Johnston was further swayed by presentations by Israeli farmers and scientists on new approaches to agriculture with novel irrigation technologies and crop management.[54] With an understanding that unused water was needlessly flowing to the sea and thereby being wasted, Johnston agreed to significantly increase Israel's share of the water so that Israel could make productive use of it.

Johnston succeeded in getting the water technocrats in each Arab country to recognize his revised plan as the basis for a fair allocation of the Jordan River for each party's use. None of the Arab nations was worse off for the compromise, but it was a win for Israel, and, at last, it was a green light for the country's most ambitious water project.

Simcha Blass's Greek Tragedy

In retrospect, every large infrastructure project seems obvious or inevitable. The cost, the sacrifice, and the risk of failure are minimized or forgotten. But Israel was a poor country, still managing the twin burdens of absorbing a large number of immigrants and defending its vulnerable borders from attack and infiltration. It took courage to look far ahead and to see water needs that would not be met by current supplies. If most elected officials defer decisions of this cost, complexity, and risk of failure, Israel's leaders accepted the challenge, possibly because the National Water Carrier had been part of the national consciousness since the White Paper was announced in May 1939.

The water pipeline from Tel Aviv's Yarkon River to the northern Negev, Simcha Blass's Phase Two project, opened in July 1955. American Jewry donated two-thirds of the money for the project with the rest raised by the government of Israel selling bonds (also mostly to American Jews). With that new water, fifty thousand acres of desert land were brought under cultivation. The opening ceremony included prayers of thanksgiving, and singers and dancers from all of Israel's leading theaters. Representatives from seventeen American cities and the governor of New York, Averell Harriman, also attended.[55]

Almost immediately, planning for the National Water Carrier, as Phase Three had come to be called, was to begin. It would bring water from the north of the country to the Negev in the south while also integrating the Yarkon River water brought to the south in Phase Two. Unique engineering challenges had to be overcome. Unlike the Yarkon-Negev line which ran along the sandy coast and was comparatively easy to plan and to construct, for the National Water Carrier, a gigantic underground plumbing system through rocky terrain had to be created to make it invulnerable to attack by hostile parties and, like any pipeline, durable enough to last for many decades.

Israel is a tiny country, often compared to New Jersey in size, but it has a wide variety of climates and altitudes. The national water infrastructure needed to function flawlessly at sea level, but also at seven hundred feet below sea level at the Sea of Galilee and at nearly three thousand feet above sea level in Jerusalem. It would also have to perform in wet, chilly winters and dry, sizzling deserts.

Nearly every part of the country would have to endure significant excavations to make way for the new pipes, pumps, and valves.[56] The inconvenience was to be significant, but every citizen, Jew or Arab, would soon experience the benefits of that inconvenience.

For Simcha Blass, this culmination of his life's work ended up having the feel of a Greek tragedy.[57] In the early 1950s, Blass had left Mekorot to serve as the new Israeli government's special representative for water issues, the most important issue for Blass being the negotiation with Eisenhower's ambassador, Eric Johnston. With that job not taxing all of Blass's abilities, and Blass wanting to spend more time planning the National Water Carrier, a government-owned water-planning company called TAHAL was created around him. It generated dozens of studies and proposals for water planning.

Blass had long assumed that when the day came to build the National Water Carrier, he would supervise the planning and construction of it. Instead, the decision was made to split the tasks and to give the construction responsibility to Mekorot, Blass's former company that he had helped to start with Levi Eshkol in 1937. Rather than accept a key role in only part of seeing the National Water Carrier come to life, Blass quit his government positions and went home to wait for the call telling him that he was right after all. The call never came. Ben-Gurion and others tried to convince him to return to the planning post, but Blass refused.[58]

A National Transformation

The National Water Carrier would prove to be far more than a pipeline of extraordinary cost and complexity. The country's new system not only improved water reliability, access, and quality overnight, it also served as a great inspiration for the young nation. Whether landing a man on the moon or rebuilding after a terrible hurricane, large infrastructure projects that are completed on time and on budget give the larger public a feeling of civic pride and enhance national identity. They also provide a widespread sense that other communal challenges can be overcome, and can unify a country. With Israel, a nation cobbled together with immigrants from more than one hundred countries, the National Water Carrier did that and more.

The pipeline was also a massive public works project for the growing nation. For several years of its construction in the early 1960s, thousands of people were digging, welding, pipefitting, or otherwise working on the new water system each day.[59] To give a sense of the scope and expense of the project, on a per capita basis, adjusted for inflation, Israel spent six times more building the National Water Carrier than the US did building the Panama Canal, which, when constructed, "was the most expensive public works project in American history."[60] On a per capita basis, Israel's National Water Carrier cost far more than such iconic US public works as the Hoover Dam or the Golden Gate Bridge.

The National Water Carrier enabled the Negev desert to fulfill Ben-Gurion's pledge that Israel would make the desert bloom. With the capacity to pump more than 120 billion gallons of water through the network, there was now a lot of water available to cultivate crops of many kinds in the arid sands of the south. Many of the arriving immigrants needing homes and professions found their way to new communities in the Negev where they became farmers.

The map of the country also changed. Until the National Water Carrier opened, the desert began just south of Rehovot, an agricultural settlement a short drive from Tel Aviv. The newly flowing water allowed the small country to settle further south from Rehovot fifty miles or more to points south of Beersheba. Today, Beersheba is a dynamic, exciting city and the regional capital of the Negev. Without the National Water Carrier, the country could not have pushed the limits of the desert and seen large numbers of people settle there.[61]

The success of the National Water Carrier definitively proved the British bureaucrats and economists wrong about the limits of population growth. While Israel's water success isn't all due to the National Water Carrier, the planning, respect for technology, determination, and risk taking that has produced Israel's mastery over weather and its abundance in water began with its planning for its national water system. From a country that could barely feed itself when it had a much smaller population, today it is not only self-sufficient in fruits, vegetables, dairy, and poultry, it also exports billions of dollars of high-quality, water-intensive produce each year.[62]

The largest part of those exports are a result of the desert cultivation, research in plant science, breeding, and genetics that all took on a new force in Israel, all areas in which Israel enjoys a leadership position in scientific research today, following the completion of the national water system. The availability of water in desert land allowed Israel's scientists—many of them immigrants—to think of new ways of utilizing that land. Ben-Gurion was proven correct: The worthless, barren desert turned out to be highly productive and valuable.

Although the environmental movement and the changes it inspired were still many years off, the ability to access this northern water took pressure off wells drilled along the coast. The wells could use winter rain to replenish and less salty water from the north could be mixed with the coastal water for better health. Ultimately, the presence of an integrated national water system also allowed for rivers to be rehabilitated from being open dump sites and sewage canals to places for recreation and nature.

Starting with Blass's first plan in 1939 prepared in response to the British White Paper, thinking about water was transformed from being a local or regional concern. Whether in Blass's hands or that of his many successors, planning and usage of water thereafter was national in outlook. This contributed to the development of an Israeli identity.

The National Water Carrier proved to be more than an infrastructure project. It was also the embodiment of the idea that the interests of the nation were superior to that of any one part. Everyone would rise together. If that ideology didn't apply in practice everywhere, it certainly became and remains Israel's governing philosophy regarding water.

The National Water Carrier was opened on June 10, 1964. Because of security concerns, there was no elaborate ceremony as there had been to mark the opening of the Yarkon-Negev Pipeline a decade earlier.[63] Visitors were invited to attend a series of small events and several were given the honor of activating one part or another of the pipeline. Blass's successor in planning the National Water Carrier, Aaron Wiener,[64] was one of these guests. Walter Clay Lowdermilk toured part of the facility during a special trip to Israel.[65] There is no record of Simcha Blass having been invited to or attending the ceremonies.

Three

MANAGING A NATIONAL WATER SYSTEM

You can tell a lot about a country by the way it manages its water.
—Shimon Tal, former head of the Israel Water Commission

After the passage of the comprehensive 1959 Water Law—which centralized ownership and control of the nation's water—and the completion of the National Water Carrier, the focus in running Israel's national water system moved to implementation. While both a strong legal underpinning and a national infrastructure were essential, it would be in the actual day-to-day governance where the average Israeli would experience the country's water system.

Israel has had high-quality regulators from the start. Given the many interested stakeholders in water governance and, even more, in allocation of resources, it is remarkable that in the many decades since the passage of the Water Law, corruption has been essentially nonexistent and the public has been overwhelmingly pleased with its water regulators, even if not always with the politicians to whom those regulators have reported.

The Water Law of 1959 appointed a powerful water commissioner to develop and execute national water policy under the auspices of a

Water Council. As powerful and apolitical as the water commissioner might be, there was still an active role for oversight by the government, which is to say by a political figure. The Water Council was placed under the supervision and control of the Minister of Agriculture.[1]

Farmers in Israel—as with agriculture everywhere—are the largest consumers of water, and it made sense, at least initially, for water to fall under the purview of the Agriculture Ministry. But as Israel grew into a modern state with an advanced economy, it became ever less logical to bias the outcome of how water should be allocated by having water policy tied to farm policy. Water, of course, was of special interest to farmers, but not exclusively to them.

Many government ministries began to stake claims to some part of the water equation. The Water Council (renamed the Water Commission) came to be under the administrative control of the Ministry of Infrastructure, but lots of other Cabinet ministers were making demands. Some of these were for good policy reasons, but in the evolving bureaucratic gridlock, politics or conflicting policy goals began to create turf wars. The 1959 law's goal of having water policies solely in the interests of the people of Israel at times became subservient to the interests of politicians.

A roll call of the different parts of the government claiming a piece of water governance gives a feel for the scope of the administrative problem. The Finance Ministry set water prices, except for the price paid by farmers. The Agriculture Ministry set that price. (The Interior Ministry played a role in setting household prices.) Sewage and its treatment were under the control of both the Ministry of Infrastructure and the Ministry of Environmental Protection. The Ministry of Health and the Ministry of Environmental Protection both had input into the criteria for water quality and safety. The Interior Ministry, aside from its role in setting domestic prices, controlled the distribution of water within municipalities. The Ministry of Justice was involved in adjudication of water disputes. The Defense Ministry had oversight involving security of water resources in the West Bank. The Foreign Ministry was the address for water-resource sharing with the Kingdom of Jordan. The Knesset Finance Committee also had oversight.[2]

One keen observer, David Pargament, said, "Imagine there was a decision to regulate trees, but that one government ministry had control over the leaves, another over the branches, another the bark, another the trunk, another the roots and yet one more, the shade of the tree. That's what happened here."[3]

By the early 2000s, pressure grew to untie the knot that wove all of these Cabinet ministries and ministers together. In an act of apparent selflessness, a few political leaders decided to push for a change in what was clearly the nation's interest, if not that of the politicians and the powerful groups who already had a seat at the table.

In 2006, following a report from a highly regarded parliamentary investigative committee urging systemic changes, the 1959 Water Law was amended.[4] The Water Commission was renamed the Israel Water Authority, and it was given real authority. Power was transferred from the political level to a technocratic one.[5] With politics taken out of the decision-making process, the newly empowered entity could make decisions without fear of being overridden by politicians who wanted to score points with voters or simply accumulate power.

"Price Was the Most Effective Incentive of All"

From the first days of the state, careful use of water was a core principle of civil life. Whether in the home or farm, Israelis prided themselves on being careful with the water they used and in developing technologies—such as drip irrigation—to be ever more mindful with its use. Every few years when a drought would hit the region, Israelis accepted the idea that they would need to make an extra effort to conserve. But the idea that Israelis could not be pushed to conserve more water soon came up against a real world test. In 2008, the Water Authority announced that everyone would have to pay the real price for the water they were using.

The reason for the price rise wasn't with conservation exclusively in mind. Rather, the water regulators wanted to maximize spending on water infrastructure, both existing and new. The promise to the public was that water fees would henceforth be spent exclusively on the nation's water needs, with nothing diverted to help balance other parts of municipal or national budgets.

As with taxpayers everywhere, the price rise didn't go over well. "People here understand that water is precious, but they still don't understand why they need to pay for it," says a senior official of the Israel Water Authority. "They see the rain and think that water is free. And they are right. *That* water is free. But safe, reliable, always available water is not free and cannot be free. Building infrastructure to get clean water to your home isn't free, and treating sewage so no one gets sick from it isn't free, and developing desalinated water plants to bring us through a drought isn't free."[6]

Before the price of water was increased, charges mostly reflected the pumping cost of getting the water to people's homes. Farmers didn't even pay the full price for transporting the water. It was common for exceptions to be granted from billing, and politicians regularly created subsidies for key constituents or favored projects.

Professor Uri Shani, the first head of the Water Authority, told the Cabinet ministers, "If you want to subsidize farmers or disabled people or give water to the country's neighbors, no problem. Discount or give away all you'd like. But whatever you take or allocate, the government has to reimburse the water utility for the water used." There would be no more free, cheap, or subsidized water, he told them. "Everyone would be on the same rules. Everyone pays."[7]

In all, household water prices were increased forty percent.[8] The public howled, and logically so. There was no apparent change in the water that came to their homes. For what seemed to be the same service, everyone was paying more. If infrastructure had always been the government's cost—like fixing a road—there was no clear reason why that should change vis-à-vis water.

At about the same time that the price hikes went into effect, the Water Authority took away management of all water and sewage from every municipality and created a new, apolitical system of municipal water utility corporations.[9] All water and sewage fees went to these new entities, leaving mayors angry that they had lost the use of these no-questions-asked fees that had long been used by them as they wished. If they had a shortfall in their municipal budgets, the water fees were always available as a backup.[10] Pipe maintenance was easy to defer while more urgent priorities got the attention of their citizens and voters.[11]

The Water Authority wanted the fifty-five new local water companies to be focused on fixing leaks, improving service, serving as incubators for new technologies, and thinking about how to save water or expenses. All of the water fees would now be spent on those goals, along with having adequate funding for building out the national water infrastructure.

While mayors had had a perverse incentive to spend as little as possible on fixing water problems as all of the unspent water fees could then be used for other municipal projects, the new water corporations had to spend all of those fees on water projects or face fines from the Water Authority. Previously, leaks were handled largely as they are around the world, often permitted to fester until an emergency was at hand. Tearing up a street makes mayors less popular, and water lost to leaks had come with no charge. If one of the new water corporations missed its goal in reducing leaks, it would be sanctioned by the Water Authority.[12] Now, if a mayor wanted the parks in his city to be watered every night, he could do so, but would have to pay for it out of his municipal budget. There would be no more "free" water for public parks.[13]

Homeowners weren't the only ones paying higher fees in the new structure. Farmers, too, were told that higher prices would be coming. Because of the longer lead times in switching crops and the hardship of a sudden price rise, a schedule was negotiated with the farmers to phase in the price increases for them. They, too, were unhappy, but were given comfort by a promise from the Water Authority that they would henceforth receive ample water once they started paying the real price. In past droughts, farmers would see their water allocations cut, and they were assured that going forward they could get all of the water they wanted to buy.[14]

The effect of introducing real pricing for farms and homes almost immediately changed usage levels. With no rationing or limit on supply, real pricing induced consumers to cut their use of household water by sixteen percent. Farmers didn't need a phased-in, multiyear step-up pricing schedule to give them time to transition to new crops. They began changing their water-use patterns in the first growing season after the announcement was made.[15]

"For the few years before the price mechanism was used," Shimon Tal, the former water commissioner, says, "we were in the middle of a terrible regional drought. The Water Commission had an ongoing and aggressive consumer education campaign on why everyone had to save water. It was a real success. Consumer usage dropped eight percent. Then we used price as an incentive. Almost overnight, consumers found ways to save nearly *double* the amount of water they had saved because of our years-long education campaign. It turned out that price was the most effective incentive of all."[16]

Cities as Labs for Innovation

The municipal water corporations turned out to be better stewards of the water of Israel's cities and towns than the mayors had been. When control of water and sewage was taken away from the mayors, there was one overarching goal: to reduce municipal leaks and unaccounted-for water use. The Water Authority was sure that the pipes wouldn't stop leaking until more was spent on infrastructure and greater focus given to innovation. If some world-renowned cities were losing forty percent or more of their water to leaks,[17] it didn't matter to the Water Authority that Israel, in 2006, was losing about sixteen percent.[18] From the Authority's perspective, it was still unacceptably high.

"Think about it this way," Abraham Tenne, the Water Authority's desalination expert, says. "We are spending more than four hundred million dollars on one new desalination plant. If we can cut our national water losses by a few percentage points, the amount of water we add is equal to what a new desalination plant will produce."[19]

Even when starting with a water-saving mind-set, people respond to the incentives they are given. They can always do better.

By 2013, lost municipal water fell to under eleven percent—a saving of nearly nine billion gallons previously lost each year. The success emboldened the Water Authority, which then set a new target of seven percent loss to leaks.[20] The success also inspired many of the water corporations to adopt some of the entrepreneurialism that the Water Authority expected of them.

Utilities are rarely known for risk taking or cutting-edge innovation. The Water Authority wanted to change that culture and to use Israel's cities as laboratories for new ideas in water. Inventors were invited to pitch concepts to the utilities as if the utilities were high-tech companies.

Nir Barlev was, until recently, the head of the Ra'anana Water Corporation, one of the new municipal water utilities. He has a deep, rich voice from his first career in opera. He then studied environmental science and went from the stage to sewage, which culminated with him becoming among the most respected heads of a municipal water company. One of the things he liked most about his job was the way in which the citizens of Ra'anana, a bedroom community not far from Tel Aviv, took part in helping to reduce water usage.

"We weren't responsible for watering the local parks. That is still the role of the Ra'anana city government," he says. "But if the sprinkler in one of the parks was spraying water onto a path, people—lots of people—called in to us to report it. And if someone spotted a leak anywhere in town, even before there is a puddle, we received thousands of calls."[21]

In a city of just over seventy-five thousand people, the claim of "thousands" of calls was used just as a figure of speech, but it gives an order of magnitude of civic involvement in trying to prevent water loss. Watering of private lawns in Ra'anana is way down, and remodeling household gardens to use little or no water is way up. Municipal buildings and city parks never had to pay for their water in the old structure; now they do. No surprise, but both public and private water use has dropped significantly—nearly thirty percent overall in this one city.

Aside from an engaged citizenry, Barlev also gives credit to the enhanced use of technology made possible by a government program. Subsidies of up to seventy percent of cost are given to local water utilities when they use new technologies of potential high impact. "The world's water crisis can only be solved with smarter use of the water we have," says Barlev. "Israeli tech companies have changed the world in computers, mobile phones, health care, and other areas. So, why not in water?"[22]

One major innovation adopted by Barlev during his tenure is Ra'anana's universal use of Distant Meter Reading (DMR) technology. Barlev describes it as a cell phone married to your home's water meter that makes a call every four hours to report on your water usage.

"Of course, we saved on not having a meter reader come to everyone's home," says Barlev, "but the real value was in the data transmitted." Working with a joint venture between IBM and an Israeli tech firm named Miltel, DMR utilizes a "consumption fingerprint" for each of the twenty-seven thousand water meters in the Ra'anana area. The system uses the same analytics that credit card companies use in trying to detect credit card fraud. If a home, business, government office, or farm is suddenly outside of its profile, the municipal utility assumes it could be because of a leak. "Nearly one in five households and businesses have suspicious water activity each year," says Barlev. "Most of the time, it is innocent, like someone filling up a boiler. But when it was a leak, we almost always knew it before the person we alerted."

As a result of this still ongoing rapid response, leaks don't go undetected for months until a water bill seems impossibly high. They sometimes only go for a few hours. "The consumer is grateful that we saved them from a large water bill or damage to their property, and the city gets to further reduce its water lost to leaks." Although the national figure of water lost to leaks is an already low eleven percent, Ra'anana loses only six percent of its water.

Ra'anana was the first of the fifty-five municipal water corporations to make use of DMR, but now several others have begun using it. "I can flatly predict that in ten years, nearly everyone in Israel will be using DMR, and within twenty years, it will be in general use around the world," Barlev says.[23]

If Ra'anana is a fairly young city with still relatively new pipes, Jerusalem has a water system that goes back hundreds of years and a history to the dawn of time. In fact, the municipal water corporation there is called Hagihon, a reference to a siege of the ancient city of Jerusalem that was broken twenty-nine hundred years ago by the construction of a tunnel to the Gihon spring.

Hagihon started as a pilot project in 1996 and this head start over other municipalities may account for its operating at a sophisticated level of service. Every pipe in the large system serving Israel's biggest city and its environs has an ID card with a profile and a leak history. Using robotic cameras, the insides of Jerusalem's sewer pipes are checked to be sure there are no cracks that would allow raw sewage to leak into the ground. Long before they become problematic, water and sewage pipes are replaced, exactly the way the Water Authority hopes all of the now well-funded water corporations will act. Despite the many parts of the Jerusalem water system that go back to the pre-State era—and a few even to the Ottoman period—water leaks are only thirteen percent in Israel's capital with many of the city's modern sectors being at six percent.[24]

Zohar Yinon, Hagihon's CEO, is ready to take on a larger responsibility than Jerusalem, and already handles some Jerusalem suburbs. But he wants more than a bigger geographic footprint. He wants his utility to be a special kind of lab for innovation, also as the Water Authority had hoped.

"I not only want to try all kinds of innovations here, I also want Israeli innovators to use us to beta test their ideas," Yinon says. "We have every kind of condition here in this city, from desert to alpine, with altitudes going as high as about eight hundred meters [about twenty-five hundred feet]. We have ancient water systems side-by-side with modern ones. We have religious communities here that don't want us to excavate what may be graveyards from long ago and we have archaeologists who demand we reroute our pipes to preserve areas for future exploration. Yet, we have to deliver high-quality water on demand to everyone who lives here. If I can help a company to develop a new idea in water, it is good for me, good for them, good for Israel, and, when they bring their innovation to other countries, good for the world."[25]

Levi Eshkol, pictured here in 1947, was one of the founding fathers of the State of Israel. He served as Israel's third prime minister (1963–1969) and led the country during the 1967 Six-Day War. Even so, Eshkol's greatest legacy may be his leadership in helping to develop the nation's water infrastructure, including his role in cofounding Mekorot, Israel's national water company, in 1937. (Kluger Zoltan/Israel Government Press Office)

Israel water visionary Simcha Blass was the key figure in every important Israeli water planning or engineering decision from the early 1930s until the mid-1950s, when a turf dispute caused him to abruptly quit his powerful government job. He left behind a transformed national water profile with implications even for today. While semiretired, Blass invented drip irrigation, ultimately revolutionizing irrigation worldwide.

While on a U.S. Department of Agriculture trip to Palestine in 1938, U.S. soil scientist Walter Clay Lowdermilk became enchanted with the soil reclamation and water management techniques employed by the Zionist pioneers there. He thereafter proposed it as a model for economic development in the Middle East and in arid regions generally. Lowdermilk and his wife became devoted to the cause of a Jewish national home. He is shown here in 1953 delivering a radio broadcast from Israel. (David Eldan/Israel Government Press Office)

PALESTINE
LAND OF
PROMISE

BY
WALTER CLAY LOWDERMILK
Assistant Chief of the Soil Conservation Service
of the United States

WITH A FOREWORD BY
SIR JOHN RUSSELL, D.Sc., F.R.S.
Chairman of Agriculture Committee Interallied Post-War Requirements Bureau: formerly Director of the Rothamsted Experimental Station and of the Imperial Bureau of Soil Science, and President of the International Society of Soil Science.

WITH 16 PLATES · 4/6

Walter Clay Lowdermilk's 1944 *Palestine, Land of Promise* was both a tribute to the Zionist revival of Palestine and a blueprint for how water could be managed there to assure millions of immigrants that they could live water-secure lives. The book became a bestseller and went through eleven printings. Although Lowdermilk's plan was not implemented, it served as a leading argument against British limitations on Jewish immigration to what was to become Israel. (Westher Hess)

Israeli expertise in deep drilling for water was initially developed to support settlement and farming in outlying areas. Wells were drilled to supply the "Eleven Points" collective farms in the northern Negev Desert with adequate water, first at Kibbutz Nir Am, shown here in 1947. Later, wells in Israel would be drilled to depths of as much as one mile. (Kluger Zoltan/Israel Government Press Office)

Establishing Jewish farms in the Negev Desert became an essential part of the Zionist plan for statehood. Utilizing the water found at Kibbutz Nir Am, a network using World War II-era British pipes was built that provided water to outlying farms. Shown here in 1947 receiving a shipment of pipes amid the desert sands, Kibbutz Hatzerim was connected to the "Champagne Pipeline," called that because of the very high price of the pipes but said to be worth it because it permitted the development of agriculture in the northern Negev. (Kibbutz Hatzerim Archive)

Israel's future prime minister Golda Meir, left, is shown in conversation with Walter Clay Lowdermilk and his wife, Inez, in Jerusalem in 1956. Despite the privations and rationing of food at the time, the Lowdermilks decided to live in Israel. Meir and the Lowdermilks shared an interest in helping the world's poorest people and nations. Meir, while serving as foreign minister, established a program providing assistance in many countries, especially in Africa, that continues today. (David Rubinger/Israel Government Press Office)

Israel's National Water Carrier was a marvel of planning and design that also required the invention of new engineering techniques. Chief engineer Aaron Wiener is pictured here in 1957—with his young daughter Ruti in the background—presenting the National Water Carrier plan, a Robin Hood infrastructure project that brought water from the relatively water-rich north to farms in the water-poor south. Wiener would go on to build TAHAL, a government water-planning division, into a leading water-planning and engineering agency for less developed countries. (Aaron Wiener Family)

Construction of the National Water Carrier was the most expensive infrastructure project, per capita, in Israel's history. It was also one of the most extensive, bisecting the country from 1959 to 1964 with excavation, construction, and laying of massive pipes. On an inspection tour, Israeli Prime Minister David Ben-Gurion, center, and Aaron Wiener, right of Ben-Gurion, are dwarfed by one of the National Water Carrier's main conduits. (Aaron Wiener Family)

The National Water Carrier was a large public works project employing thousands of workers at any given time. To speed the work, crews often excavated from opposite sides. Here, workers tunneling through rock reach for laborers on the other side. Despite the massive scope of the project, construction took about five years. When completed, the system would carry billions of gallons of water to the Negev, fulfilling Ben-Gurion's pledge that Israel would make the desert bloom. (Daniel Rosenblum/Mekorot)

Water infrastructure is built to last. Israel's National Water Carrier had the additional burden of having to be impervious to attack. It also had to be equally effective in a wide range of altitudes and in weather patterns as varied as cold winters in the Galilee and Jerusalem and sizzling summers in the Negev Desert. Shown here is a massive pipe placed in a rocky formation that is about to be connected. Workers pose for a camera in this early 1960s photo. (Daniel Rosenblum/Mekorot)

After years of inconvenience and sacrifice, the National Water Carrier was opened in June 1964. It almost immediately transformed the country, a legacy still felt today. Despite the feeling of national pride, because of security concerns there was no large public celebration to mark its opening. TAHAL's Aaron Wiener was among those given the honor of turning a switch or valve to begin the flow of water. (Daniel Rosenblum/Mekorot)

Beginning with elementary school, the standard Israeli curriculum includes lessons about conservation—that it's everyone's responsibility. As part of that training, students are given tips on how to bathe and brush their teeth in a water-efficient way. By the time schoolchildren become adults, saving water is ingrained as an important part of everyday life. This classroom poster from 1960 reminds students that "It Is a Pity to Waste Even a Drop." (Ze'ev Lipman)

Israel posthumously honored Prime Minister Levi Eshkol in 1985 on its lowest-denominated paper currency, the then-new five New Shekel bill. Recognizing his legacy developing the country's water infrastructure, in general, and the National Water Carrier, in particular, the designers of the note included an artistic rendering of a pipe with water flowing from the hills of Israel's north to the sands in the south.

Many water-themed subjects have been depicted on Israeli postage stamps, from ancient aqueducts built in the time of the biblical kings to contemporary water technology. To mark the seventieth anniversary of Mekorot in 2007, a postage stamp was issued honoring the pre-state water company that grew into the national water utility. The stamp calls out a range of Mekorot functions, including water purification, deep well drilling, pipeline construction, and rain-cloud seeding.

When a prolonged drought hit Israel in the late 2000s, the Israel Water Authority launched a national advertising campaign to urge even lower consumption of water. Nine Israeli celebrities appeared in the "Israel is Drying Out" series of ads, in which their famous faces began to crack like parched soil. One in the series, a 2009 television commercial featuring Israeli supermodel Bar Refaeli, was widely discussed and became symbolic of the necessity to conserve water. (Israel Water Authority)

Reminders to conserve water are found in public places throughout Israel, from office bathrooms to schools to water fountains to beach showers. All send the same message that every citizen has a role to play in safeguarding the country's water. This waterfront metal sign, designed to evoke the blue stripes of the Israeli flag but with a water faucet in the place of the Jewish star, uses a command Hebrew tense with the instruction: "Save water!" (Seth M. Siegel)

PART II

THE TRANSFORMATION

Four

REVOLUTION(S) ON THE FARM

One day in the middle thirties, I happened to pass near the fence of Abraham Lobzowski's house and saw there a tree some ten meters high, much taller than any other tree along this fence.
—Simcha Blass

THERE ARE FEW PEOPLE who reinvent themselves at age fifty-nine and have a second career as important as their first. Israel's "Water Man," Simcha Blass, was one of them.

Following Blass's principled, if miscalculated, resignation from the project to plan and build Israel's National Water Carrier, he went from being the most important person in water in Israel for more than twenty years to being an ordinary citizen who followed the country's water developments in the morning newspaper. After a few years of mostly marking time in self-imposed semiretirement, in 1959, Blass returned to an idea that had first occurred to him more than twenty-five years earlier.

While visiting a farm to supervise the drilling of a well, Blass, then a young water engineer, had noticed an anomaly in a row of trees planted along a fence; one was much taller than the others. Blass knew that they were all of the same species, likely planted at the same time, and they all had the same soil, sun, climate, and rainfall. Why, Blass wondered, was that one tree in the row growing so robustly?

Walking around the tree, Blass found a tiny leak in a metal irrigation pipe near the base. He suspected that these small, but steady, drops of water were going to the tree's roots and were the likely cause of its superior growth. The image of that tree stayed with him. "I became busy with other plans," he wrote many years later, "but the drop of water that grew a gigantic tree refused to leave me. It stayed trapped and sleeping in my heart." Decades passed, and, with his life in disarray and needing a project, Blass decided to explore whether that one outsized tree was an oddity or the precursor to a completely new way of irrigating trees and field crops.[1]

Ideally, crops and orchards grow with no irrigation at all. When there is consistent and seasonably predictable rainfall, no human intervention with irrigation is needed. But frequently, where the crops are is where the rain isn't. Even if there is adequate rain, it may come at the wrong time or not consistently enough. Whenever the shortfall of rain might sabotage a farmer's plans, he needs to augment the rain with irrigation—water originating in lakes, rivers, reservoirs, or aquifers and brought to the crops.

At the time that Blass began his exploration, the most common form of irrigation was flood irrigation. Fields or their furrows would be inundated with water, or, in the case of orchards, trenches would be dug around the base of the tree, and the trench would be flooded.

Flood irrigation has been in effect since the dawn of civilization in the Middle East where water from the Nile River in Egypt or the Tigris-Euphrates River system in ancient Iraq would be diverted through gravity-fed canals to irrigate large tracts of land.[2] Around the world today, flood irrigation is still widely in use, even on farms in regions without abundant water.[3] The wastefulness of flood irrigation is especially vivid in places far removed from a water source where enormous effort and expense is needed to bring the water to the crops only to see the majority of the flood-irrigation water evaporating or draining uselessly into the soil before it can be absorbed by the roots. In general, more than fifty percent of flood-irrigation water is wasted.[4]

Despite Israel being semiarid, as Blass began his experiments in the late 1950s, flood irrigation had also been a common form of irrigation there.[5] Water for agriculture then consumed over seventy percent of Israel's total water usage,[6] as is still typical for most countries today.[7] If, he imagined, agricultural water could be reduced by even a few percentage points through smarter irrigation, more food could be grown or the extra water could be available for household use for the country's fast-growing population.

The alternative to flood irrigation at that time—different varieties of sprinkler irrigation—suffered from similar problems. Anyone who has seen a lawn sprinkler in action, especially when there is even a gentle breeze or the spray nozzle is not perfectly directed, knows that much of the water ends up on a sidewalk or far from its intended target. Some spots get too much water, others not enough. The same happens in a field. Plus, the airborne spray is aloft just long enough for much of the water to evaporate before it hits the ground. In all, experts estimate that about one-third of the water is lost with sprinkler irrigation.[8]

Irrigating a plant drop by drop limits evaporation and delivers the water that the plant needs directly at its roots. The water savings are significant—only four percent of the water is lost to evaporation or unnecessary absorption into the soil.

While the idea of drip irrigation seems simple, an everyday example illustrates why it is an engineering challenge of monumental complexity. Windowsill plants are usually watered in a form similar to flood or sprinkler irrigation. Taking a watering can and pouring a lot of water into a flowerpot is akin to flood irrigation. Most of the water will either evaporate or drain out at the bottom. Using a misting bottle and aiming the nozzle at the plant's leaves and roots is something like sprinkler irrigation. Less water is wasted, but still a large amount is lost.

Mimicking drip irrigation with that windowsill plant, though, would require someone standing over the flowerpot with an eyedropper and dripping water targeted at the roots. But even that doesn't capture the complexity because much of drip irrigation is subsurface, with the drippers buried in the soil adjacent to the roots. That imaginary eyedropper would have to be planted a few inches below the top of the soil. But then the dropper could get stopped up with dirt or the plant's roots could grow into the dropper, as happened in one of Blass's earliest field trials, nearly dooming the project from the start.

To imagine the challenge facing Blass, that single windowsill plant needs to be multiplied by the long rows in a field, each consisting of hundreds of plants, all needing an equal amount of water at the exact same moment. The water must be delivered in what could be a wide range of temperatures and weather conditions. Since water pressure at the end of the line would be lower than at the start—and because it is common for one part of the field to be higher than another—Blass and those who followed him had to devise a means to equalize pressure through the entire line and to overcome the effects of gravity.

Had Blass decided that day in 1933 when he observed the row of trees to immediately begin work on drip irrigation, he would have been unlikely to develop a device that worked consistently. Blass's initial assumption, even in 1959, was that he would be working with metal pipes similar to the one he found with the leak dripping on to the root of the larger tree. But the passage of time served Blass well.

During World War II, there had been a revolution in material science. Plastic began to be used as a substitute for traditional materials like metal and glass. Plastic could not only serve as an inexpensive substitute for metal pipes, but it could be molded in very precise dimensions down to a fraction of a millimeter.

After several years of "experimentation and struggle,"[9] as Blass termed it, with different materials, delivery systems, types of trees and plants, and with varied water quality, Blass made two discoveries. First, as he had hoped, regardless of the location where he did his experiment in Israel and regardless of the type of tree or field plant, drip irrigation used far less water than was used by flood or sprinkler irrigation on adjacent test areas. On average, drip irrigation saved fifty to sixty percent of the water customarily used.

But the second—entirely serendipitous—discovery would prove to be even more important than the amount of water saved: In every experiment Blass conducted, the yield from crops watered with drip irrigation was higher than with other known irrigation techniques. With no additional acreage to be planted, the enhanced harvest was akin to getting free crops with no extra water used. Even farms in water-rich areas would benefit from irrigating plants with drip irrigation. It was an invention with the potential to change the world of agriculture.

The Dubious Professors

Maybe every breakthrough idea has its share of naysayers. If drip irrigation had come from a more mild-mannered inventor or a less notorious one than the temperamental Blass, perhaps it would have received a better reception. Or perhaps the claims of what drip irrigation could do stretched the boundaries of credulity; irrigation hadn't changed much in thousands of years. But rather than being welcomed and celebrated for the revolutionary idea that drip irrigation was, Blass mostly failed to get the support of people in academia, government, agriculture, or business who could appreciate what he had invented.

In the early 1960s, Blass presented his findings to scholars and practitioners at the agriculture faculty of Hebrew University, then and now the country's preeminent institution for soil science, irrigation, and agronomy. They mostly scorned the idea. Blass had the further misfortune that the junior faculty member who ran a series of experiments that proved drip irrigation's efficacy in saving water, and especially in producing a higher yield, wasn't taken seriously by the faculty because he lacked advanced academic credentials and suffered from a writing style deemed insufficiently scientific.[10]

Using his government contacts, Blass then turned to the Ministry of

Agriculture and asked its extension service to run a series of experiments with drip irrigation in an almond orchard. The experiment quickly came to an end when the roots migrated into the drippers, blocking the flow of water and killing all of the water-deprived trees. Drip irrigation nearly died with them.

Lucky for drip irrigation's future, a Ministry of Agriculture extension service field officer, Yehuda Zohar, suggested that the experiment be run a second time, but with the drippers sitting at the base of the trees rather than in the soil. The second set of trees thrived. Water was saved and, again, yield greatly enhanced. The experiment's success gave Blass the (short-lived) confidence to begin looking for a business partner to bring drip irrigation to market. He made ten presentations and was turned down every time.[11]

Netafim Is Born

Yet another fortunate intervention saved Blass and drip irrigation.

At the same time that Blass was being rejected by would-be partners and manufacturers, several of the country's socialist collective farms began thinking that they needed a manufacturing business to balance their agricultural activities. One of these was Kibbutz Hatzerim, which was among the eleven settlements established in the night after Yom Kippur 1946 to solidify the pre-State claims to the Negev desert.[12] Ironically, it had been Blass who arranged the water hookup to the kibbutz, and it was now, in part, the lack of adequate water supplies that led this socialist farm to seek a nonagricultural business.

Blass was dismissive of the kibbutz's interest in his invention. On the one hand, he had nearly given up hope of selling it and thought that only "a complete idiot" would be interested in it. And on the other, he was quite sure that the inexperienced kibbutz personnel would fail to properly manufacture it. But the member of the kibbutz responsible for finding a manufacturing business, Uri Werber, was neither put off by Blass's sometimes abrasive style nor his wavering faith in the effectiveness of his product.[13]

Werber had had a friendly relationship with Yehuda Zohar, the Ministry of Agriculture field officer who had convinced him that, Blass's own misgivings notwithstanding, drip irrigation had enormous potential. Werber's persistence paid off when, a few months later, Blass sold the rights to his invention to the kibbutz for twenty percent of the stock in the company to be formed and also a small royalty on sales to be paid to him and to his son, who was his business partner.[14]

Werber asked one of his Kibbutz Hatzerim colleagues to come up with a name for the company and they settled on Netafim, from the Hebrew word meaning "to drip." The company began operations in January 1966.

The original market for Netafim's drip-irrigation equipment was other farms in Israel, and the product was a near overnight success. Exports of the equipment soon followed, and growth overseas was strong from the start. But the success brought with it a problem. Because the Kibbutz Hatzerim members were true to their socialist ideology, they refused to have employees and insisted on doing every part of the manufacturing and sales by themselves. This put a limit on how much Netafim could produce.

By 1974, when the Kibbutz Hatzerim members working at Netafim could no longer handle all of the business opportunities in Israel and around the world, the kibbutz decided to give away—for free to another kibbutz—exclusive territorial distribution rights in parts of Israel and in important countries around the world. In 1979, with growth still overwhelming these socialist businesspeople, Kibbutz Hatzerim and its first partner, Kibbutz Magal, shared their business opportunity—again, for free—with Kibbutz Yiftach. All three became the joint owners of Netafim.[15]

The idea of giving away large parts of a profitable, fast-growing company and simultaneously diluting management control seems to defy logic. But to Ruth Keren, a Kibbutz Hatzerim veteran who now serves as the head archivist of the kibbutz's rich trove of historical material and whose deceased husband coined the name Netafim, the change in ownership made perfect sense. "We lived by strict principles, and one of them was that we will only do what we can do ourselves," she said. "We would not hire workers. Since we couldn't do it ourselves, we decided to give it away."[16]

Even with its two partners, Netafim still could not fill demand. Three unrelated kibbutzim set up drip-irrigation companies of their own in the 1970s and went into competition with Netafim.[17] One Israeli inventor with his own approach to drip irrigation set up a company in California and entered into a business arrangement with a Greek entity named Eurodrip. Realizing that the world's many Islamic countries honoring the Arab boycott of Israel were a rich market for drip-irrigation products and that these countries would not openly buy from an Israeli company like Netafim or its Israeli competitors at that time, Eurodrip was able to mask the product designer's national origins and to seize that opportunity.[18]

All of these Israeli and Israel-related companies are still active in one form or another, but each of them was ultimately wooed by the lure of capitalism. The two kibbutz competitors of Netafim sold their drip-irrigation businesses to large international companies.[19] Likewise, the two kibbutz partners of Kibbutz Hatzerim got rich from the opportunity given to them, and all three partners ended up selling large stakes to private equity investors. European private equity fund Permira now owns over sixty percent of Netafim, and Kibbutz Hatzerim owns most of the rest.[20] The several Israeli-originated drip-irrigation companies still dominate the industry worldwide, which enjoys annual sales now of over $2.5 billion with Netafim, the largest, selling about $800 million of that.[21]

Blass and his son both became quite well-off from their respective stakes in Netafim. They shared in the profits of all three Netafim kibbutz enterprises and got royalties, too. Over time, both he and his son sold back their ownership stakes and royalty participations to Netafim for large lump sum payments. Blass lived the rest of his life at a level of comfort not possible on an Israeli government pension.[22]

Kibbutz Hatzerim also became enriched by drip irrigation. A photo of the socialist farm taken a few days after its founding evokes a desert moonscape with a single tree breaking a rocky, sandy horizon. Today, on that same spot, the shared kibbutz and Netafim campus is a vista of low buildings, lawns intersected by walkways, and an abundance of trees. A visitor noted that the kibbutz evokes a small village,[23] and the one thousand people who live there seemingly enjoy a comfortable middle-class life.[24] A few steps from the kibbutz, Netafim's factory for its drippers runs several shifts each day with many of the employees—Negev Bedouin, Russian, and Ethiopian immigrants, and native-born Israelis—commuting to work from around the Negev region.

Better Than Rain

Simcha Blass may have been the creator of drip irrigation, but Rafi Mehoudar is its most prolific inventor. From a family that has lived in Jerusalem for twelve generations, Mehoudar attended the Technion, Israel's elite institute of engineering and technology. While still an undergraduate, he began inventing, even winning a competition with a device to improve sprinkler irrigation.

In 1972, soon after Mehoudar's military service and his graduation from the Technion, Netafim contacted him, hoping he would join the company's R&D department.[25] A lifelong city dweller, Mehoudar had never heard of Netafim, by then a staple on Israeli farms. Mehoudar declined to become an employee, but agreed to work with Netafim on a royalty basis, a wise decision for him given what followed. In the decades since, he has developed dozens of innovations and upgrades that have vastly improved drip irrigation, including keeping drip levels uniform throughout a field, even if on a mountainside, and reengineering Blass's device with more effective, precision-molded fittings.[26]

Before Mehoudar became involved with drip irrigation, two principles had already been established. First, use of drip irrigation can save as much as seventy percent of the water that would otherwise be used for irrigating the crop. That number isn't always so high, but forty percent water savings are now routine.[27]

Second, drip irrigation will produce a larger harvest and usually a higher-quality one, as well. Regardless of growing conditions or the salinity of the water, drip irrigation nearly always produces a larger crop than does flood- or sprinkler-irrigated crops in a comparable environment. Harvests of double or more are now standard. In some recent controlled-environment studies in the Netherlands, state-of-the-art drip-irrigation equipment caused increases of up to 550 percent over open-field irrigation—while saving forty percent of the water.[28]

For countries where water is not in short supply, like the Netherlands, water savings are less important, at least for now. Saving on the energy cost of pumping the water to the field is a plus in both reducing the use of carbon fuels and in lower operating costs for the farmer. But the most significant benefit in these locales is that drip irrigation offers extraordinary opportunities, especially in greenhouses, for growing more. In a low-margin, high-risk industry like farming, the prospect of more for less is an important safeguard against the inevitable reversals growers will experience.[29]

As to why drip-irrigated crops do so much better, Mehoudar has a possible answer. "By giving the plants too much water, which is what we are doing with flood and sprinkler irrigation, the roots get flooded and deprived of oxygen," Mehoudar says. "This stresses the plant. Then we give them no water for some period, and this stresses them in a different way. We do this over and over again during the growing cycle. On the other hand, when we drip water regularly for the plant, we keep it calm and let it do what it does best."[30]

Since flood irrigation mimics, in a way, periods of rain and no rain, it would seem that drip irrigation isn't just a superior irrigation technique, but better—and more consistent—than rain itself.[31] Of course, drip-irrigation equipment is more expensive than rain, which may be unacceptable for those willing to accept the unpredictability of rain and the lower crop yields it produces.

Saving the World from Algae Infiltration

Aside from the benefits on water saving and yield, another Israeli-created innovation in drip irrigation can save lakes and streams from algae infiltration, a significant environmental hazard. When rainwater falls on overfertilized fields, huge quantities of the millions of tons of fertilizer used worldwide each year get washed into lakes and other freshwater sources through the ordinary runoff process. The phosphorus or nitrogen in the fertilizer serves as a dynamic food source for naturally occurring algae. When enhanced with a few days of warm temperatures, the algae go through explosive growth. The blue-green algal blooms often seen in lakes deplete the oxygen in the water, rapidly killing plants and fish. The water smells foul, and can only be utilized for drinking, washing, agriculture, or recreation after an expensive cleaning process.

The half million people in the greater Toledo, Ohio, area experienced a huge algal bloom in the summer of 2014. Toledo is on Lake Erie, one of the Great Lakes and among the largest freshwater resources in the world. Toledo's residents were told to not drink the water at risk of intestinal ailments and also not to bathe with it to avoid breaking out in rashes caused by the toxins emitted by the algae. Although adjacent to Lake Erie, these half million people were reliant on bottled water, and functionally out of usable water until the algae crisis could be resolved.[32] In recent years, around the world (including in Israel), thousands of blue-green algal blooms have been found in water supplies due to agricultural runoff from fertilizer or animal waste.

Drip irrigation is a solution to the fertilizer-incited algae colonies. Instead of randomly distributing fertilizer on fields, drip-irrigated plants often receive a mixture of water and water-soluble fertilizer in a process called fertigation, a new word coined by combining *fert*ilizer and irr*igation*. Just as a lot less water is used with drip-irrigated plants, much less fertilizer is also used. This saves the farmer the expense of all of that extra fertilizer and spares society from its deleterious effects, including environmental disasters to clean up later. It also keeps freshwater resources like Lake Erie drinkable at a time when freshwater is growing scarcer and more precious. The fertigated water is dripped right at the root and is taken up by the plant. Almost no trace of fertilizer is left behind to be washed into waterways with the next storm—or to leach into the soil and pollute the underground water supply for the next generation.

Fertigation logically led to nutrigation. Global population growth is occurring just as a shortage of quality farmland is being felt worldwide. With billions of new mouths to feed and less quality land on which to grow food, nutrigation addresses the problem by permitting farmers to grow their crops on subpar land—or even in desert sand—that lacks the necessary nutrients provided by rich soil. Just as fertigation brings fertilizer to the plant, nutrigation provides all of the nutrients that traditional crops would receive from the soil along with the plant's drip-irrigation water.

Danny Ariel is a specialist in agriculture in developing countries. "What wheat is to the Western world, rice is to the developing countries of Asia," Ariel says. "Rice is grown on riverbanks or floodplains. The population in these countries is growing but there isn't enough floodplain land available for growing more rice. But you can also grow upland from the flood area if you are using drip irrigation. The rice farmer could still grow what he grows by traditional means on the floodplain and, at the same time, also grow a second crop of rice adjacent to his existing rice fields."[33]

"With nutrigation," says Rafi Mehoudar, the prolific drip-irrigation inventor, "crops can grow anywhere. The sand of a desert can be used to hold the plant in place and the nutrigated water can do the rest. The job of the soil is no longer about providing nutrients. Now, it can be just to anchor the root as it waits to be fed."[34]

And innovations in saving water with drip irrigation continue. "As good as drip irrigation is at saving water—and it is very good—there is still a big opportunity," says Professor Uri Shani. Shani began his career in water as a professor of soil science at Hebrew University, and then served as the first head of the Israel Water Authority. He followed that by playing an integral role in putting together the Red Sea–Dead Sea project that will transport desalinated water from the south of the Kingdom of Jordan as a prelude to sharing water among Israel, Jordan, and the Palestinian Authority. He now spends his days as an inventor and businessman in agriculture. In partnership with Netafim, he recently developed the next stage in drip irrigation, an inexpensive device that sits in the soil adjacent to the roots and sends a signal when the plant needs water or nutrition.

"Currently, with drip irrigation," he says, "water is dripped at the root at intervals when we believe the plant needs it, but we may be guessing wrong on the timing. The plant takes in the water, but much of that water then evaporates out of the plant."[35] This "irrigation-on-demand" system, Shani says, referring to the new system he developed, is "drip irrigation that listens to our customers," with those "customers" being the roots of billions of plants.[36]

Breeding Plants to Grow with Less Water

The farmers in pre-State Israel relied upon local Arab merchants to provide them with seeds for their vegetables and other field crops. In 1939, during a period of deteriorating Arab-Jewish relations, the Palestinian Arab leadership declared a ban on sales of seeds and other farm products to Jewish farmers. In reaction, the kibbutzim and other Jewish farms banded together and created a cooperative to serve local farmers with seeds of consistent quality. The co-op was named Hazera, from the Hebrew word for seed.[37]

At the time, seed production and breeding companies everywhere in the world largely developed seeds for their own local climate and soil conditions. The world's first seed-breeding company, Vilmorin, was founded in France in 1742, and although it is a global leader today, it was a local French company for most of its history. Similarly, soon after its creation, Hazera began working on a variety of new strains of seeds to address specific problems like local insects and plant diseases, but especially on a search for seeds that would perform well in water-stressed environments, a singular concern for the Jewish farmers who were its customers. If a plant could thrive with less water, it would not only be able to withstand periodic droughts, but it would also put less demand on water resources.[38]

Following independence in 1948, with the creation of many new farms and the arrival of millions of new immigrants, Hazera scrambled, at first, to keep up with the demand for its seeds, but by 1959, Hazera began exporting its surplus seed inventory to countries with climates similar to Israel's. Soon thereafter, it evolved into a global business with offices around the world and seeds customized for the local growing conditions of its customers. Hazera also incrementally expanded its research facilities.

As with high-tech, semiconductors, biotech, and cyber security in which Israeli R&D is seen as a key resource for other global companies, Israel is now a leader in plant research. Its laboratories design solutions for farmers in many countries, but it has a high degree of awareness of the needs of Israeli farmers. Graduates of Israel's universities—especially Hebrew University, the Technion, and Ben-Gurion University—populate the laboratory benches at Hazera, as well as Evogene, a newer Israeli company similarly focused on plant genetics. Global seed and agricultural companies like Monsanto, DuPont, Syngenta, and Bayer have R&D centers in Israel or have made acquisitions or set up joint ventures with Israeli seed companies.[39]

Although Israeli seed-breeding companies conduct research in both traditional breeding and genetically modified (GMO) varieties for their many international customers, no farmer in Israel ever uses GMO seeds. This isn't because of a scientific rejection of it, but rather due to market sensitivity. With a broad consumer distrust of GMO products in Europe and with many European customers for Israeli produce, a decision was made to only use seeds in Israel bred by traditional methods.[40]

For the domestic market, the Israeli seed breeders have found two important ways to save large amounts of water needed to grow crops. First, plants are bred to be as water efficient as possible. "In looking at the plant," says Dr. Moshe Bar, an Israeli seed-breeding expert, "we think about what are the essential elements and what are not. It takes water to grow every part of the plant, so it makes no sense to encourage any more evapotranspiration," the process by which water evaporates out of a plant, "than is necessary."[41]

For example, Israeli seed breeders developed a new kind of short-stalked wheat for Israeli—and now other—growers. "The stalk adds nothing to the wheat, so why waste water growing it?" asks Dr. Shoshan Haran, former senior plant researcher and manager at Hazera, now utilizing her breeding skills as the head of Fair Planet, an Israeli NGO she founded. Fair Planet specializes in coming up with unique, localized seeds for conditions faced by impoverished African farmers.

Likewise, Israeli plant geneticists came up with a breed of tomato for Israeli farms with few leaves and with the tomatoes growing closely together. "We left enough leaves on the tomato plant to protect them from the sun and bred the tomatoes to grow together to get the plant as compact as possible. This saves a lot of water, because you don't have to expend water growing leaves or longer vines," Dr. Haran says. "With our focus on the yield—that is the amount of tomatoes and their weight—we looked to minimize other parts of the plant."[42]

The seed breeders working in Israel also customized the root structure of some plants to save on what they came to think of as needlessly long roots. When utilizing flood irrigation—which includes an interlude of no irrigation—plants grow long roots to follow the water down. With drip irrigation and its steady dripping of water, with no loss of product quality and with meaningful water savings, roots of the same plant species can be bred to be about one-third the length of the same species utilizing flood irrigation.[43]

Seeds for Salty Water

Aside from rethinking the elements of the plant, Israeli seed breeders also came up with a revolutionary, and counterintuitive, idea of developing plants that thrive when grown on brackish water, the previously useless salty water found in abundance under the Negev desert's sands and throughout the Middle East. By growing a significant amount of the country's fruits and vegetables using otherwise undrinkable water—and using it to spur a multi-billion-dollar agricultural export industry—Israel is able to improve its citizens' diets and enhance its economy, without putting a strain on its freshwater resources.

When Simcha Blass first showed his drip-irrigation model to the Israeli professors at Hebrew University, they told him that even if he could overcome technical problems, which they doubted was possible, the device would "only succeed if the irrigation could be carried out with entirely salt-free water, *aqua destillata*, for if the water contained any amount of chlorides, the soil would become saline" and be ruined.[44] Since all of Israel's natural water has a fairly high salt content, this may have been another way for the academics to tell him that his idea had no prospect of success.

The professors were, of course, wrong about drip irrigation, but they were also wrong about the potential of irrigating with the naturally salt-laden Israeli drinking water. Israeli plant geneticists have taken it a step further, creating melons, peppers, tomatoes, eggplants, and other fruits and vegetables that thrive on diluted brackish water. Now, researchers at Ben-Gurion University and Hazera are developing melons that can grow with even saltier water that would further reduce the amount of freshwater needed to dilute the brackish water used for irrigation. If successful, this is likely to set off development of other salt-absorbing fruits and vegetables.[45]

As plants absorb the salty water, there is a change in the plant's cell structure. The amount of water in the cell declines, but the natural sugars increase. This produces sweeter fruits and vegetables with a better texture. "For now, the only negative, if it is a negative, is that the produce is slightly smaller," says Moshe Bar, the Israeli seed breeder. "Everything tastes better, and the marketplace has noticed." Drip-irrigated produce grown with diluted or slightly desalinated brackish water now has a wide following in Israel, and also in export markets in Europe and Asia.[46]

Shoshan Haran sees Israel as a model for the water-stressed world that is coming. "The focus for Israeli farmers and breeders has been on water and water scarcity for a very long time. No one anywhere has more experience," she says. "Farmers all over the world will soon need crops that can thrive with limited rain or even in drought. From long experience, Israel knows how to grow food under these conditions."

The best place in Israel to grow crops today is in the desert, Haran says. "I know it is irrational, but it is so because of the kinds of seeds and type of irrigation we use," she says. "With climate change, this is an opportunity for Israel to share these new seeds and Israeli know-how with the rest of the world. Drip irrigation and these special seeds are important in Israel today. Soon they will be important around the world."[47]

Finding "Trillions of Gallons of Water"

In getting access to the brackish water that helps drive its desert agriculture, Israel has developed a geological and hydrological expertise in identifying where that water is, how much is likely to be found, and how best to get to it.

The brackish water found in the Arava desert in Israel's far south is all secreted in nonrenewable aquifers. Because of the thick rock layer insulating them, rain does not percolate into these aquifers to renew them. The water found there is from a prior geologic era, and, fittingly, these nonrenewable sources are also called fossil aquifers. As with oil extraction, once this highly mineralized water is pumped out, the water is depleted. The caverns of fossil water may be vast and, with controlled pumping, the water can last for decades. But once gone, it is gone forever.

Ami Shacham has been involved with Israel's desert agriculture and the search for fossil aquifers from the beginning. He moved to the central Arava, a part of the Negev desert, as a young man in 1959, "before there was air-conditioning," he says. "It was a very hard life. We lived in tents while housing was built. And what woman would want to pick up and raise a family here?" But Shacham managed to meet such a woman, and they raised a family, with two of their children and five grandchildren still living in the sparsely populated desert.

In his long career as the head of water-resource management in the Arava, Shacham has overseen the drilling of fifty-five wells, one nearly a mile deep, and the construction of a complex reservoir system for capturing the rainwater from the ferocious flash floods that roar through the desert in winter. All of the billions of gallons of water he has extracted from these many wells were very salty, needing either dilution or brackish water desalination, a specialty of Mekorot. Four hundred privately owned farms have been set up in the Arava during Shacham's tenure, with seventy-five percent of the produce grown sold for export.[48]

Professor Arie Issar, now retired from Ben-Gurion University of the Negev, was the very first to argue in the 1950s that the sandy wasteland of the Arava desert sat atop water that could be used for agriculture and development. "People laughed when we began drilling in the Arava," he says. "But look at the central Arava today with its fishponds and with its fields of crops. In Israel, there is a sea of water underneath the soil. From here to the Sahara desert, there are [trillions of gallons of] water. If you can drill thousands of meters to get oil, why can't you drill far less to get water for agriculture?"[49]

Those trillions of gallons of non-potable, brackish water under the sands of the Middle East have always been considered worthless. But Shacham and the Arava farming enterprise demonstrate that, while also adding to the variety of their diet, Israel's neighbors could use the brackish water under their sands to add to or even transform their agriculture-based economies.

A Moral Challenge for an Affluent World

The benefits of drip irrigation are both technical and social. Drip irrigation saves water, enhances yield, and helps reduce use of carbon fuels because it requires less energy than other irrigation methods. It also adds to the supply of arable land for growing crops, reduces degradation of aquifers, slows or stops the algal bloom problem, and reverses creeping deserts. Further, it is an important tool in grappling with world hunger and the political upheaval that often comes with it. Drip irrigation also helps socially by reducing poverty through building greater capacity, and improving the status of women who are less obliged to spend their days hauling water. "Drip irrigation holistically addresses the point where all of these challenges intersect," says Naty Barak, Netafim's chief sustainability officer.[50]

Around the world today, only about five percent of the *irrigated* agricultural fields utilize drip irrigation or other micro-irrigation techniques.[51] To put that into context, less than twenty percent of the world's fields now utilize any irrigation with the rest relying, at least for now, on rainfall.[52] The penetration of advanced irrigation is sure to increase worldwide with reduced rainfalls coming at the same time as there will be more mouths to feed. But of the fields that are irrigated around the world, about eighty percent[53]—including many in the United States—still use some form of the ancient, and wasteful, flood-irrigation method.[54]

By contrast, drip irrigation is the norm in Israel where seventy-five percent of all irrigated fields can be found with drippers in or on the ground, with the rest using sprinklers. Not one farm in Israel has used flood irrigation in several decades.[55] Israel's widespread use of drip irrigation makes sense since it was invented there and was widely installed in farms in Israel before much of the world even knew of its existence. Drip-irrigation equipment can now be found in use to one degree or another in 110 countries, but no country makes use of drip irrigation as comprehensively as does Israel.[56]

While global drip-irrigation adoption levels are still low today, there is no doubt that the number of farms and fields utilizing drip irrigation will grow, and significantly so, in the next decade. Water is becoming too scarce. The need to enhance yields of crops is too dramatic. Fertilizer is expensive and, in any event, its use must be reduced. Much, if not most, of the world's high-quality arable land is already under cultivation and the next millions of acres will come from lower-quality soil—even desert soil, as is the case in Israel.

Fittingly, the most dramatic increases in the use of drip irrigation have occurred in China and India, where the latter now leads the world with over five million acres utilizing the technique.[57] The largest drip irrigation company in India is Jain Irrigation, a large Indian agriculture conglomerate that acquired the Israeli company NaanDan.[58] Netafim is the second largest drip-irrigation company in India.[59]

In both rich and poor countries, most governments now widely subsidize the price of water to the point of making it seem to be free.[60] These subsidies come from more than the general government coffers. Since there is a finite amount governments will spend on supplying water to farms and homes, those water subsidies crowd out other government spending, such as for quality assurance testing, new infrastructure, and/or new technology. The pocket picked for functionally free water isn't always the taxpayers' pocket, even if that is also often the case. More often, it is from a better water future not pursued.

As water becomes scarcer, price will become the most effective tool for managing household demand and, even more important, for utilizing market forces to functionally ration agricultural water. Once there is a cost to farmers for the water they use—as is the case in Israel—farmers will have an incentive to modernize their farms and to use all kinds of technology to preserve water and to purify marginal water. Among other changes, this will likely catalyze a broad and global transition from flood irrigation to drip irrigation.

"Israel has had a role in several revolutions in agriculture," says Technion water expert and professor emeritus Uri Shamir. "Israeli agriculture has reduced the amount of freshwater it uses by sixty percent. This was done by changes in crops utilized, irrigation techniques adopted, and the rise of technology."[61] With a price for water factored in, purchasing drip-irrigation equipment will make ever more sense to help end wasteful flood irrigation and irrational crop selection.

Separate from water being subject to market mechanisms, though, drip irrigation should be utilized in the world's most impoverished communities because it is among the best tools for reducing subsistence agriculture poverty—a moral challenge for an affluent world. Small farms using drip irrigation thrive agriculturally and these enhanced yields often provide the economic development necessary to climb out of dire poverty. With donor nations and foundations eager to improve the lot of the "bottom billion," the embrace of drip irrigation will widely improve people's lives while addressing many of the greatest challenges the world faces today.

While government subsidies generally distort the marketplace, the introduction of technology for poor farmers may be a wiser use of government resources than subsidized water.[62] India's farmers have adopted drip irrigation so broadly in large part due to state governments providing subsidies to do so—and with universally positive outcomes in food production and improved incomes for farmers. It has also enhanced fairness in distribution of water resources. For example, in a twenty-nine-thousand-acre area in Karnataka, in southwest India, the even flow of drip irrigation is being used to assure that every farmer in the system gets the same amount of water at the same time.[63]

As the growing need for irrigation will lead to the use of different sources of water (coming from desalination, reclaimed sewage of varying quality, brackish fossil water, or a mixture of all three), drip irrigation works well with all sources of water. Different drippers have been developed for different levels of water purity; no water source is incompatible with drip irrigation.

Drip irrigation also provides a special opportunity for donors to make a difference in the lives of the world's poorest people. Since there is a cost to drip-irrigation equipment, it is a logical project for the many Western philanthropists involved in venture philanthropy and also for the many new charitable organizations giving micro-loans for projects in Indian and African agriculture. In recognition that most poor farmers in less developed countries have no energy source to run their drip-irrigation equipment, a gravity-fed version has been invited. Micro-lenders will find many of the world's hundreds of millions of subsistence farmers eager to take on this debt—and to repay it with the profits from their larger harvests—once the farmers have been shown how drip irrigation can enhance their lives. It is a truism that it serves people better in the long term to teach them to fish rather than to give them fish.

Drip irrigation, Naty Barak, Netafim's chief sustainability officer, believes, is among the best ways to improve the world. "Access to water is a human right, and we should think of it being as important as freedom of speech or freedom from persecution and other human rights," he says. "It may even be more important, because without water, we can't live for more than a few days."

"Drip irrigation itself doesn't give people more water for drinking or for sanitation," he says. "But worldwide, agriculture uses about seventy percent of our water. Only ten percent is used for drinking, cooking, and keeping clean. If a country can reduce its agriculture water use by just fifteen percent—an easy goal with drip irrigation—that extra water would more than double what people have available to them. Using drip irrigation, Israel has done that and more."

"The world should think of Israel as a laboratory, but also as an inspiration," Barak says. "If we can do it out here, in the middle of a desert, anyone can do it."[64]

Five

TURNING WASTE INTO WATER

There is no shortage of water. The world is full of water,
but most of it is dirty. The challenge is to clean it.
—Sandra Shapira,
Israeli water executive

IN 1950, LESS THAN TWO YEARS after Israel's independence, government officials began discussing the then extreme idea of using the country's sewage to irrigate some of Israel's crops.[1] Although the idea was soon rejected due to health and "aesthetic" concerns, the conversation had begun.[2]

Facing a perpetual need for new water resources, Israeli government officials and farmers overcame their initial objections and, over several decades, have built an agricultural economy and a national wastewater infrastructure to make use of that sewage. No other country makes the reuse of its sewage a national priority as does Israel. Over 85 percent of the nation's sewage is reused.[3] If in the US, as in most of the world today, sewage reuse is negligible, in the coming water-constrained world, it is a near certainty that everyone will soon be turning to highly treated sewage as an essential new source of water for agricultural and other uses.

Although sewage was once mostly a nuisance, even a source of pollution, it is now seen in Israel for its value as a parallel water system in an arid region. As such, it is now seen as a treasured national resource. If anything, Israeli farmers now wish there were more of it.

Sewage is made up of everything that goes down the sink, shower, bathtub, or toilet. It may also include the rain that falls into the storm drains found in most municipal streets. Ideally, sewage is disposed of in a separate collection and distribution system running alongside, but never touching, the freshwater network.

At its best, all of the collected wastewater is treated before it is discharged into a river or, better yet, reused. But in some countries, the raw sewage is only directed away from the source of the waste and discharged—untreated—into lakes and rivers, creating health and environmental risks to those bodies of water, and threats to aquifers below.

Throughout human history, people lived in proximity to their own waste and often contracted diseases from it. Only after Dr. John Snow, a British anesthesiologist, identified a contaminated well as the source of London's 1854 cholera outbreak did the idea of separating wastewater from drinking water take hold. In 1858, in response to Snow's discovery, London began conveying sewage to a downriver section of the Thames River—far enough away that the smell would not disrupt daily life while safeguarding the cleaner upriver water to be used for drinking, washing, and other household purposes.

In the decades that followed, as Europe suffered through yet more cholera epidemics in cities that failed to separate their freshwater and sewage, it became apparent that Snow's hypothesis was correct. Simply avoiding wastewater could prevent mass death as well as providing a higher quality of life.[4] Cities everywhere began channeling untreated sewage into waterways like rivers and seas far from the inbound source of the city's drinking water. Following this one important realization, sewage treatment remained the same for nearly one hundred years.

Right after World War II, mostly in the US and Great Britain, the idea of treating sewage before dumping it took hold. The motivation wasn't a concern about pollution or a case of incipient environmentalism. Rather, it was in response to the erroneous belief that untreated sewage was causing polio just as sewage-tainted river water had once caused cholera. Although no causal link to polio was found, by the 1950s, wastewater treatment—primitive by today's standards—came to be seen as a fundamental part of municipal life. The practice grew around the globe along with postwar prosperity and the expansion of government services everywhere. Today, there are more than one hundred thousand wastewater treatment plants worldwide.[5]

Initially, sewage treatment consisted of two processes. After an initial pretreatment during which large items like trash and debris were removed from the water as it flowed through a series of screens at the entry to the treatment center, the primary treatment would begin.[6] The watery, brown-colored, foul-smelling sewage would be directed into large tanks where the heavier solid and semisolid organic matter in the water would settle on the bottom of the tank due to gravity.[7] This organic matter, or sludge, would be removed from the tank, often put in sealed packages, and disposed of in landfills. The remaining, still quite contaminated, water would then be drained into a river or ocean via a dedicated pipeline.

Soon it was determined that the dissolved organic material in the sewage that survived the primary treatment process was causing oxygen depletion in waterways, and so another process was added. After the primary treatment, a sophisticated combination of benign bacteria and lots of oxygen was added to the mix. The organic matter remaining in the sewage—human waste, food particles, skin washed off in the shower—would be consumed by these hungry, but friendly, bugs. The bacteria would get fat and heavy from this temperature-controlled, oxygenated feast, and sink to the bottom of the tank, where, as with the primary treatment, the material on the bottom would be taken away.

At this secondary treatment stage, most of the organic matter was removed, but it was still possible that viruses and other toxic material remained in the wastewater. In addition, at this secondary stage, an odor would usually linger, providing a reminder, in case anyone needed one, that the secondary treated sewage still wasn't safe or pure. Nevertheless, the water was of higher quality than before, and, as with the primary treated sewage, it was discharged into rivers and oceans.[8]

As environmental concerns grew worldwide in the 1970s, those countries and municipalities that could afford to do so added a third, or tertiary, level of treatment. This process would disinfect the wastewater with chlorine, ultraviolet radiation, or other means before being safely discharged into a body of water.[9] Clean as the water was now, treated sewage continued to be seen mostly as a bother and a societal cost to be gotten rid of, like trash. It was rarely seen as an opportunity.

Like the rest of the world, initially, Israel dumped its sewage, untreated. A dedicated pipeline was built for the wastewater generated by inhabitants in Tel Aviv and the other coastal cities. The pipeline ended about a half mile into the Mediterranean where the sewage would be spewed about ten to fifteen feet below the water's surface, hoping, no doubt, the tide would carry the waste out or onto the seabed. Israel's inland cities would rely upon nearby rivers to convey their sewage to the Mediterranean. Despite the best hopes of the engineers who designed the system, with a shift of the waves, the cascades of sewage would, at times, flow back to the shoreline, befouling Israel's beaches, and compromising Israel's embryonic tourism industry.[10]

By 1956, the greater Tel Aviv area's seven municipalities—jointly called the Dan Region—comprised about a third of Israel's national population, and an even larger percentage of its sewage. A decision was made to aggregate all of the Dan Region's wastewater and to send it via a huge pipe to a mostly uninhabited area about eight miles south of Tel Aviv. There, the region's sewage would be treated at a facility named Shafdan, an acronym for the Hebrew "Dan Region Sewage." Due to budget and engineering challenges, the project took far longer to complete than first imagined, with the facility finally servicing all of the municipalities only in 1973.[11]

As complex as building the wastewater treatment plant turned out to be, there was the hope, but no certainty, that when the facility was completed some of the Shafdan's treated wastewater might be used for agriculture.[12] What no one could have expected was how completely Shafdan would change Israel's water profile, agriculture industry, and the economic development of the Negev desert region.

A Completely New Source of Water

Five miles south of Shafdan and a short distance inland from the Mediterranean Sea, there are sandy dunes that sit over an aquifer nearly three hundred feet below. In thinking about a new way to treat sewage, and in planning a radical departure from the then common practice of dumping it into the sea, Israeli government geologists and hydrologists in the late 1950s began to wonder if the fine sand above the aquifer could serve as an additional filter to clean the still tainted, secondary-treated wastewater.[13] Like treatment plants typical of that period, Shafdan was set up to treat sewage in both primary and secondary processes, but the facility lacked the capability to bring it to the higher level of safety and purity that tertiary treatment provides.

There was risk in finding out if the sand filter would work. If the partially treated sewage that was percolated into the aquifer made the six-month to one-year journey through the sand with viruses and hazardous microparticles still intact, the underground reservoir would be at risk of contamination and the water from the aquifer might become unusable for drinking or bathing. But if the sand filtration did work, the dunes could provide a large-scale, chemical-free solution for processing the large daily volume of Shafdan's treated sewage.[14]

The Israeli engineers saw several benefits if the sand could work as a treatment medium. First, there would be no need to build a large, new wastewater treatment facility. Second, once purified by the sand, the huge volume of water produced by the process could be stored in the aquifer and pumped out, as needed, saving on the need to build a storage reservoir. But most important, the treated water could be used for irrigation.

Every part of the process required bold thinking and often significant expense. Getting the seven cities—later joined by eighteen other cities and towns—to agree to consolidate their sewage in a regional center and building a mega wastewater-treatment plant was, by itself, a substantial task.[15] The decision to attempt to use sand, a process that came to be called Sand Aquifer Treatment, or SAT, to obtain tertiary-quality reclaimed water challenged conventional scientific and engineering wisdom. Perhaps most striking, the decision to convert a freshwater aquifer into a specialized treated-wastewater one was an educated risk of the kind that governments and utilities rarely choose to make. That this risk was taken in a country with little margin for losing existing water resources made it yet more remarkable.

An equal challenge for the engineers working on Shafdan was to ensure that none of the treated water in the aquifer would spill over into any of the freshwater aquifers nearby. The volume of water in this reserved area would have to be continuously monitored and special wells would need to be drilled on its periphery for the purpose of observing and monitoring the aquifer now filled with treated water. It was one thing to ruin a single aquifer; the country could not afford to lose a series of them.[16]

While the Shafdan planning and tests were taking place, a powerful senior official at the Ministry of Agriculture, David Yogev, began making the argument that even without the SAT water, secondary-treated water from both Shafdan and other treatment facilities constructed elsewhere in Israel could be used by farmers.[17] Two other ministries were not convinced, and they voiced concern.

The Ministry of Health was worried about crops absorbing toxic matter from the less-than-perfect secondary-treated wastewater. If the plants did so, the scientists at the ministry wanted to be sure the toxic matter could not be transmitted to people. Likewise, if a crop irrigated with reclaimed wastewater was used for animal feed, there needed to be certainty that nothing harmful would resurface in eggs, milk, or the meat of the animal. In response to these concerns, and after extensive testing, it was agreed that only certain types of crops—initially, inedible ones like cotton—could be irrigated with wastewater that had gone through less-than-complete treatment.

The Ministry of Environmental Protection had a different set of concerns. Even if the treated wastewater would be used on an industrial crop that would never be eaten, the Ministry's scientists wanted to be sure there would be no effect on Israel's wells or other groundwater. If a crop was irrigated with water still containing potentially hazardous microorganisms, they wanted to know if invisible toxins might percolate into the soil and ruin an aquifer that might be below the crops. Without caution, careless use of treated sewage could compromise much of Israel's essential groundwater. The ministry developed a detailed map showing exactly where treated wastewater of differing quality could be used. Strict guidelines were developed for usage anywhere an aquifer could be put at risk.[18] Although farmers would need a special permit for using treated water of any kind, many officials had long expressed concern about trusting farmers to stick to the exhaustive irrigation guidelines.[19]

To everyone's great relief, SAT proved to be a perfect tertiary treatment. The six-month to one-year descent through the sand into the aquifer stripped away all impurities, and the water was of superb quality. The hopes of the Ministry of Agriculture to find a new source of water for agriculture were fulfilled, and the concerns of the other ministries about contamination proved to be unfounded. Over time, farmers—with some education, financial incentives, and desensitization—would come to appreciate, and ultimately to rely upon, this new resource.

Thereafter, a dedicated pipeline—with a six-foot diameter and a fifty-mile length—was constructed from the Shafdan reservoir to the Negev, providing a new source of irrigation for the farmers there.[20] At first, there were limitations as to the crops on which the Shafdan water could be used, but after years of testing, Shafdan water was permitted to be employed as freely as freshwater. Today, water from Shafdan can be used for anything except drinking.[21]

Shafdan was, and is, two facilities in one. It was a wastewater treatment plant—the largest and most technologically advanced in the Middle East—that addressed growing environmental and public concerns about the pollution of Israel's rivers and seashore. But Shafdan also played another role by helping to rethink Israel's sewage treatment and the role of reclaimed water in agriculture. After Shafdan, every municipality in Israel came to see its sewage as a resource for the betterment of the country, and even as a tool for keeping water infrastructure expenses down. And after Shafdan, farming was changed. Farmers would no longer need to scramble for water allocations.

Today, Israel treats about ninety-five percent of its sewage, with the rest going through septic systems. "Nothing Israel is doing in capturing such a large amount of its sewage for treatment is very different from other advanced countries around the world," says Avi Aharoni, Mekorot's director of wastewater and reused water. "What is different, what is extraordinary, is the degree to which Israel then takes that treated water and makes productive use of it."[22]

Utilizing a separate national collection and distribution infrastructure used only for Israel's treated wastewater, about eighty-five percent of that treated water is available to farmers for their crops. A portion of Israel's reclaimed wastewater is also used to increase the water volume of its rivers, enhancing their well-being. There has also been some planning done to begin using tertiary-treated wastewater for fighting forest fires. Israel's farms still make use of a great deal of freshwater, but treated sewage now makes up about a third of the national water used in agriculture, or about twenty percent of all water used for all purposes. In all, more than one hundred billion gallons are reused each year.[23] The national goal is to be reusing even more of the country's sewage, with a plan to be at ninety percent in another few years.[24] By comparison, Spain is second in the world in reuse of reclaimed water with around twenty-five percent, even if most affluent countries like the US reuse less than ten percent.[25] It is only a question of time before highly purified water is used for crops everywhere in a way similar to how it is used in Israel today.

Reclaimed water transformed Israel's water profile. And no less than with the role of drip irrigation and specially bred, drought-resistant seeds, the comprehensively treated sewage water changed the agricultural landscape, permitting Israel to feed itself and to be an agricultural exporter of significance—whether in years of abundant rain or scarcity.[26]

Wastewater Infrastructure Saves Israel's Agriculture

Capturing rain is a surprisingly expensive and unsanitary method for acquiring new water supplies. The last time a raindrop is pure is before it hits the ground. As raindrops amass and begin to flow, they pick up a variety of pollutants, most especially grease and soot from car and truck exhausts. Compounding this quality problem, in the case of Israel, a large amount of sand and grit, blown from beaches and desert, are also carried along with the rivulets of rain.

Even if rainwater can be cleaned at a reasonable price, it is still an unreliable source of water. In some years there will be too much rain for the available storage capacity, and more concerning, in many years there may be less rain than farmers need. Reclaimed sewage is more reliable because it isn't dependent on the vagaries of climate and rainfall, and even with all of the infrastructure required to develop it, reclaimed sewage is ultimately cheaper.[27]

Israel still uses its existing storm water capture facilities to trap rainwater and store it. But these were mostly built in the 1980s, and Israel hasn't developed a new one in recent years.[28] Perhaps more important than the actual rainwater captured each year, these facilities gave Israel the opportunity to develop sophistication and know-how in water storage, especially in building a network of reservoirs, but this time for treated wastewater. The several hundred reservoirs that began being built around Israel in 1995 to hold treated wastewater are now an essential part of a complex, multifaceted system that holds the treated water, called effluent, coming from Shafdan and from the nation's other wastewater-treatment plants.

"By the 1980s," says Yossi Schreiber, a senior executive at the Jewish National Fund (JNF) in Israel, "it became clear that without recycled water, agriculture, as it was then practiced, would have to come to an end in Israel." A decision was made to go beyond Shafdan and to attempt to have every city, village, town, and farm treat its sewage and then transport it for reuse in agriculture. "But," says Schreiber, "it couldn't happen immediately. A process had to be created which included developing new skills, building a new national infrastructure, and finding the money to pay for a big, expensive new system."[29]

Aside from the sewage pipes from homes to the treatment plants all over the country, new pipelines had to be laid to bring the treated water to storage facilities, and then another new set of pipes would be needed to convey that water to the farmers in the field. The water couldn't arrive just any time, but had to be available when needed. Further, since the country's wastewater-treatment plants didn't produce water of a uniform quality, water from each treatment plant had to be matched with reservoirs that held water for specific crops or irrigation locations.

Sewage is created every day in approximately the same amount, but farmers need water for their crops only at certain times of the year. Because of seasonable variability as to when crops are planted and need to be irrigated, the national network of reclaimed-water reservoirs was seen as a key part of the country's parallel water infrastructure.

Partnering with Israel and its farmers, JNF's US branch and branches in other Jewish Diaspora countries raised funds from local donors to assist with between thirty percent and half of the total reservoir construction cost in Israel. The remainder of the money needed to build the reservoirs was provided by the Israeli government, the Israeli office of JNF, and farmers' water co-ops.

Even though Israel is an increasingly affluent country, the upfront costs of the periodic resculpting of the nation's infrastructure is beyond what the country can bear, especially given its ongoing defense burden. While developing countries around the world get assistance in building out their water infrastructure from governmental organizations like the United States Agency for International Development (USAID) and comparable aid organizations in Europe, US assistance for Israel is specifically targeted for Israel's military and security needs and may not be used for water infrastructure.[30]

The global support from JNF for water, rural development, forestry, and environmental projects in Israel makes a large difference in the quality of life there, especially in what is called the periphery, or those areas outside of the heavily populated center of the country. The build-out of Israel's still unfinished reclaimed water system would have taken years longer to create without the efforts of JNF and the generosity of its supporters, all of which is ongoing.

With proper maintenance, these reservoirs have a lifetime of decades, but they are expensive. For example, an individual reservoir through which 800 million gallons can flow in a year costs about ten million dollars to build. There are already 230 reservoirs of different storage capacities in Israel, and JNF plans to build forty more as soon as matching government funding—now mostly focused on building desalination infrastructure—becomes available.[31]

Unlike Rain, Sewage Is Consistent, Reliable, and Predictable

Aside from all of the expense and disruption of construction, and before finalizing the national reclaimed-water project, the government needed to be sure there would be a customer for all of this new water. "Farmers had to be educated about the benefits of using treated water," JNF's Yossi Schreiber says. "At first, they were very resistant in doing so."[32]

Farmers had been getting allocations of freshwater from the Ministry of Agriculture's Water Planning Department since the first days of the state. No one wanted reclaimed water if they could be getting freshwater, or what is called "sweet" water in Hebrew. But when farmers were told that they would get a twenty percent extra allocation of treated water for every unit of their freshwater allocation they did not use, the perpetually water-craving growers began to sign up. The treated water was also offered to them at a sharply reduced price, giving them even more incentive to switch. And, due to the source of much of the treated water, the water they were getting was rich in nitrogen, saving the farmers on their fertilizer costs.

But the best incentive to convert to treated water wasn't a direct financial one. Unlike the volume of rain, which changes from year to year, the amount of sewage being turned into reclaimed water is consistent, reliable, and predictable. When farmers were promised that—unlike the fluctuating annual freshwater allocation they each had to endure—their annual treated-water allotment would be locked in, farmers' commitment to reclaimed water was assured.

If they ever had reservations about using treated sewage on their fields and in their orchards, Israel's farmers have none today. "The quality of the reclaimed water is now so high and the price is so low that farmers are constantly asking us to increase their allocation of treated water," says Dr. Taniv Rophe, the Ministry of Agriculture official in charge of water allocation to farmers.[33]

Following the laws of supply and demand, Israel has begun raising the price of its treated water, just as it earlier put farmers on notice that they would be paying full market price for freshwater. "Farmers," Rophe says, "will have to factor in the cost of water just like they factor in all other costs." Getting rid of subsidies, she says, will lead to smarter crop selection, use of innovation and technology in irrigation, and most important, more efficient use of water.[34]

Running Out of Sewage

Israel still has a problem, of sorts, with its sewage. Just as a big market for all of the treated water produced each year got developed, the supply of it began shrinking—even as the population was growing. Put simply, Israel is creating less sewage than before.

Israel has long been a water-efficient country. Children are taught in school to take short showers and to shut off the water while soaping up. There are even parts of the early-childhood hygiene curriculum that teach students to close the tap while brushing their teeth.[35] But household water savings aren't left to whim or chance. Flow restrictors are obliged to be placed on every faucet in the home.

Of even more significance, Israel was the first in the world to adopt the mandatory use of dual-flush toilets,[36] a device claimed to have been invented in Israel.[37] Plasson, an Israeli kibbutz company, had been manufacturing plastic toilets since 1973. In the 1980s, it bought a patent from a professor at Ben-Gurion University with a clever idea: variable flush volumes in toilets. The company spent several years developing it before bringing the idea to the government.

Shaul Ashkenazy, a former head of Plasson, says it was easy to convince the government it should require dual-flush toilets. "About thirty-five percent of household water use is for flushing toilets," says Ashkenazy. "With dual-flush toilets, a home can save half on most of its flushes and as much as twenty percent of its total household use of freshwater. If you multiply that by all of the homes and toilets in Israel, and by all of the reduced flushes, it comes to quite a bit of water."[38] The regulation applies to more than homes; dual-flush toilets—and not just of the Plasson version—are obligatory in every office, restaurant, hotel, or public setting in Israel. The Israel Water Authority estimates that the use of dual-flush toilets results in an annual water savings in Israel of about seventeen hundred gallons of water per person—or about 13.5 billion gallons for the nation, thanks to this single innovation.[39]

Consumer education about water conservation was long ubiquitous, even if Israel's recent abundance in water has reduced the effort and government media spending for its ad campaigns. Every Israeli, though, is familiar with widely found "Every Drop Counts" signs which are literally translated as "It is a pity to waste even a drop" and also with a series of television ads in an "Israel Is Drying Out" campaign showing Israeli celebrities' faces (including that of supermodel Bar Refaeli) becoming parched and cracked due to water shortages.

More than education or dual-flush toilets, free-market pricing of water led to a large drop in consumer usage, most especially in watering lawns and gardens, which were widely replaced with native plants, decks, and smaller areas allocated to backyard grass. Shrubs and trees in Israeli homes are nearly all linked to a low-tech version of drip irrigation to save on water. Further, most of the landscaping in public parks is irrigated with treated wastewater.

Although gardening water has no bearing on what goes down the drain inside the home, it represents a national mind-set that—even with the bounty provided by desalination—consumers prefer, at least for now, to make careful use of water inside and outside their homes.

The low-flush toilets and other water-efficient appliances, along with the constant conservation efforts, led to less water going into the country's sewage supply as a percentage of its total volume. Even if this smaller supply has come to bedevil farmers and government water allocators, the reduced volume of sewage has an economic benefit of its own beyond less wear and tear on treatment plants. With Israel's raw sewage being the least diluted, most dense, sewage found anywhere in the developed world, Israel's wastewater facilities operate at high rates of efficiency, not needing to treat excess water as is the case with the ordinarily highly diluted sewage usually found in other countries' sewage and most especially in the US.[40]

Even Reusing Toilet Paper

As is the case worldwide with sustainable development—a new subset in smart environmental policy—Israel seeks to make its sewage treatment as energy efficient as possible. In the treatment process, natural gas is produced as a by-product and is captured to help run the facility's operations. Already, this biogas powers more than sixty percent of Shafdan and other wastewater-treatment plants. The goal is to make the process energy net neutral in the next few years. That will occur both by ever lower energy use and by finding more usable energy within the sewage and during the treatment process.

The energy savings is being driven by a growing number of Israeli water-tech companies. Just as Israel is a global leader in a variety of science-based fields of innovation, the treatment of sewage has attracted an array of interesting ideas around which companies in Israel have been created.

Eytan Levy has cofounded two of the most talked-about companies in wastewater treatment. A chemical engineer trained at the Technion, Israel's MIT, Levy bubbles with excitement when he speaks about the subject. He punctuates his speech with interesting historical and scientific tidbits about sewage. His first company, Aqwise, figured out a way to use inexpensive, but durable plastic parts not much larger than Lego pieces to house the bacteria needed during the secondary, or biological, treatment process. By increasing the density of the bacteria present in the treatment tank, between double and triple the amount of sewage can be treated with the same amount of electricity used. With about two percent of total US electricity usage consumed just aerating bacteria during secondary treatment (and another two percent transporting the sewage),[41] the ability to increase secondary treatment efficiency is felt quickly in both financial and environmental dividends. As important, for budget-stretched or space-constrained communities, fewer new treatment facilities need to be built.[42]

Levy's second company, Emefcy, has a product that similarly focuses on increasing efficiency, but also on reducing the volume of activated sludge—the dead, friendly bacteria—created by wastewater treatment. His company's process tricks the bacteria into eating each other after they have finished their feast on the organic material, leaving fewer of them at the end of the treatment. This results in less by-product that needs to be transported to landfills, a meaningful energy and environmental benefit in itself.

Of even more value, the Emefcy process utilizes a new way of aerating the sewage that may reduce energy consumption by as much as ninety percent. It works best in population units of five thousand people or fewer, offering the opportunity for an additional, if unrelated, energy cost savings. "If you live in an area with a few thousand people and a golf course," says Levy, giving one example, "the sewage the population produces can be treated to excellent quality with minimal electricity." The water from the treatment process can then be used on local gardens and on the golf course without the dual-energy expenses of pumping the sewage out of the community and of pumping freshwater a long distance into the community to water its gardens and fairways. That energy savings is over and above the energy saved in efficiently aerating the sewage tank.

Apart from energy savings, another industry incubated by sewage treatment—resource recovery—has also caught the attention of Israeli innovators in wastewater. "It would be a mistake to look at sewage as all the same," says Technion professor Noah Galil, an expert in sewage treatment. "For example, household cooking grease and other oils sit on top of the sewage as it enters the treatment facility. These were initially removed with skimmers and disposed of to avoid clogging the treatment machinery. Today, that layer of oil is sold for reprocessing and used in industrial applications."[43]

Israeli entrepreneurs have asked what else is in the sewage that could be harvested and reused. "There have been attempts to remove and sell heavy metals like cadmium and selenium," says Galil, "but, so far, without success."[44]

One self-described "former tree hugger," Dr. Refael Aharon, lamented all of the trees being cut down around the world and began wondering where a replacement for the cellulosic material used in the pulp and paper industry might be found. His investigations led him to inbound raw sewage. It turns out that sewage is filled with tiny—often microscopic—bits of toilet paper, washing-machine lint, and even parts of discarded fruit and vegetables that aren't removed by bacteria or any bacterial process. In 2007, he founded a company called Applied CleanTech (ACT) that harvests the surprisingly large volume of cellulosic material found in raw sewage and coined the name "Recyllose" for it.

"With so much of wastewater solids comprised of cellulosic fibers," says Aharon, "sewage-treatment plants can be turned into mines of recycled products and reusable commodities. It is a green, unlimited, endless source for recycled pulp. By removing the cellulosic material from the wastewater-treatment plant, you increase plant capacity by thirty percent, save on energy, and need less space and cost with sludge handling."[45] ACT has sold its sewage mining product solution in several countries around the world. Aharon estimates that two-thirds of the system's value is in energy and landfill savings and a third in the value of the harvested Recyllose itself.

The Salt Hazard

With all of the good news in sewage, a subspecialty has developed that studies some potentially bad news, too. These investigators—mostly professors and government officials, but also some farmers—want to know what the impact of all of the reclaimed water in agriculture means for Israel's soil and for the health of the nation.

One concern is that Israel's natural water—that is, the water from the Sea of Galilee and from the country's aquifers—already has a relatively high salt content when it enters homes as freshwater. By the time it leaves kitchens and bathrooms, the saline level has become much higher due to all of the salt in food that gets eliminated or washed down the sink.[46] All of that blended salty water is what flows into the country's wastewater-treatment plants.

The wastewater's secondary treatment removes bacteria, and the tertiary treatment removes viruses and other microorganisms, but salt is not removed in either of those treatments. Reclaimed water is dripped onto crops in fields all over the country, with the water being taken in by the plant, but with minute quantities of salt being absorbed by the soil.[47] Farmers and others in Israel have long asked how much salt the soil can bear.

The ancient Sumerian civilization, located in what is now southern Iraq, was once among the most fertile areas of the ancient world. Its agriculture was the source of its wealth and, in turn, its power. Over hundreds of years, the monarchy developed a complex irrigation system that brought somewhat saline water into channels used to irrigate the fields. An ancient Mesopotamian tablet was found that describes how "the black fields became white" as the "broad plain was choked with salt."[48] Sumerian civilization fell.[49]

Israel's recent development of desalinated water resources—water that has had the salt removed from it—may come to the rescue. Israel's national supply of drinking water is far less saline than it was only a few years ago. With desalinated water making up more of the country's water than any other source, the use of nearly salt-free seawater results in ever less salt coming into Israel's homes in the first instance. With less salt coming in, there is less salt going out and into the wastewater.

Hormones in the Water

Aside from salt, there is now also a concern about the role of pharmaceutical products that get into the water supply. As with salt, these currently survive the wastewater-treatment process.

Over the past hundred years, as the global pharmaceutical industry has grown, doctors have come to recognize that only a small amount of medicines are absorbed by the human body. On average, about ninety percent of every pill swallowed—but never less than seventy percent—is excreted a short time later.[50]

"The [US] Food and Drug Administration," says Tel Aviv University's Professor Dror Avisar, "only tests to be sure that the medicine is safe based on primary human use. No one—not the FDA, not the EPA—tests to see what happens to the pharmaceutical residues after the medicine leaves our bodies."

Avisar is a hydrochemist. He is an expert in a still young field in which Israeli professors and laboratories have established a leadership role. These pioneering scientists are trying to figure out not just what happens to the hormone in a birth control pill once it gets passed to the wastewater-treatment plant, but also how that compound may change when it is mixed with antibiotics or with any of the thousands of different pills or personal care products that are now found in wastewater everywhere in the world. "When most people think about water safety," says Avisar, "they think of developing countries with pathogens in the water that kill babies. That is an important concern. But there may be a different cause for concern in the water of developed countries."[51]

In the past ten years, hydrochemists have become skilled at measuring pharmaceutical contaminants and residues in water, even though the quantities are extraordinarily tiny. Measurement of the compounds are found in the infinitesimal, at parts per *trillion* of volume.

Even so, Sara Elhanany, who heads the water quality division at the Israel Water Authority, says, "We are taking this very seriously. No one is avoiding it. We monitor it. But we have yet to find any relationship between the compounds in the water and any change in health for the Israelis who eat the crops irrigated with this water."[52]

Avisar doesn't see this as a uniquely Israeli problem, even if Israel has developed a self-interested academic and regulatory expertise in figuring out the potential dangers. "If an individual compound is dangerous in isolation, we don't know if it becomes dangerous or completely breaks down when exposed to sunlight or what the effect of different temperatures might be," he says. "It may also be that because we use our wastewater for irrigation, the compounds break down in the soil."[53]

Both Elhanany and Avisar, separately, take pains to say repeatedly that no one knows yet if this is a problem, but also point out if it is, it could be a significant one—and one with global implications. For example, New York City and all of the cities along the Atlantic coast in America treat their sewage and then discharge it and the pharmaceutical residues it contains into rivers that flow into the Atlantic Ocean. People eat fish caught in the Atlantic. Already, in Canada, there have been cases of fish dying from ingesting the estrogen found in birth control pills and of male fish caught that had developed female sex organs.[54]

For cities that utilize water from local rivers for their drinking water, the problem may be even worse than with ocean-harvested fish. "All along the Rhine River and other rivers in Europe," Avisar says, "cities discharge their treated sewage along with the pharmaceutical compounds." Cities downstream take in water from that same river, and use it for their drinking water before returning it to the river as treated wastewater. "By the time the same water has passed through a few cities along the river," Avisar says, "there is no question people are consuming some residual pharmaceutical products in their drinking water. We have developed the tools to measure the minute presence of these compounds and to be aware of the potential harm, but we don't know yet what advice to give. The science is still developing."[55]

Israel, says Elhanany, will do whatever it needs to do to keep both its citizens and its agriculture safe. "We have gone through the very expensive process of demanding that every new wastewater-treatment plant treat sewage to the tertiary level so that people and crops are safe," she says. "Within a few years, also at great expense, the majority of our secondary-treatment plants will be converted to tertiary."[56]

Elhanany hypothesizes a level of treatment no one considered only a few years ago. "It may be we discover that all of our wastewater has to go through a fourth treatment, or that we switch to a membrane-based treatment for all of our treated sewage," she says. Already, MemTech, a membrane company based on the invention of two professors from the Technion, has developed a means for filtering pharmaceutical molecules at the nano level from municipal wastewater.[57]

"There is no discussion whatsoever about giving up on reusing our reclaimed water. Treating wastewater and reusing it is safer for people's health than treating it and discharging it," Elhanany says. "If more treatment is needed, we will do it."[58]

From Waste to "Wow!"

Reclaimed water has had effects on Israel beyond the obvious benefit of creating a new water supply for agriculture. Without the benefit of the treated water, Israel would have had to become a dry economy, with limited or no agriculture. Fruit, vegetables, and grain would all have had to be imported.

In an arid region with limited water resources and a growing population, the reclaimed water helped to take pressure off the natural water assets. With reclaimed water as an added resource, the country can better withstand a drought without overtaxing its natural water supply.

There are other benefits. The majority of Israel's population is clustered around a very few urban centers. By having abundant water available for agriculture, the country is able to disperse from its core. Each of these outlying areas—today primarily agricultural regions—offers the potential to serve as a new zone for development and the greater dispersion of Israel's population from the crowded center of the country.

Israel is the only country in the world which has less area covered by desert today than fifty years ago.[59] A satellite photo of Israel dramatically demonstrates this, showing built-out cities throughout the country and a large green swath across the western Negev desert. Elsewhere, the problem of creeping deserts, called desertification, is creating economic and social problems in many arid countries.[60] As the desert moves in, communities move out, often creating social dislocation and exacerbating local poverty in those countries. The painstaking development of agricultural bands around Israel, especially in its arid south where all of the Shafdan water is sent, offers a promising model to other countries.[61]

If anything, Israel has succeeded too well in the reuse of its sewage. With less supply and more demand for this parallel water resource, farmers have to pay more for it. If the farmers of a generation ago received heavily subsidized freshwater, but complained that there was too little of it, this generation of Israeli farmers have all of the water they want from natural or alternative sources, but are concerned that the price they will have to pay will make their produce unaffordable in the world market.[62] Israel's water planners are betting that the cost will bring yet more efficiency to the nation's farms, and that it will, in turn, spur yet more new Israeli innovation and technology.

Israel began the treatment of its sewage to reduce pollution, and to improve the quality of its citizens' lives. Today, the country's rivers are cleaner. Israeli pollution of the Mediterranean is significantly reduced, and soon to get even more so.[63] And Israel's aquifers are at less risk of contamination. Along the way, Israel developed a parallel water supply that may not be ideal for drinking or bathing, but which can be used safely in agriculture. Regardless of climate, every country produces great amounts of sewage. Around the world, the rapidly growing number of water-stressed communities can learn from Israel's example and turn sewage from a nuisance into a precious resource.

Six

DESALINATION: SCIENCE, ENGINEERING, AND ALCHEMY

The irrigation of the desert with purified seawater will appear a dream to many, but less than any other country should Israel be afraid of dreams capable of transforming the natural order. . . . All that has been accomplished in this country is the result of dreams that have come true by virtue of vision, science, and pioneering capacity.

—David Ben-Gurion (1956)

THE ASSASSINATION OF President John F. Kennedy occurred two weeks before the Weizmann Institute's 1963 fund-raising gala in Manhattan. Kennedy had been announced as the keynote speaker and with his sudden, violent death, the event's organizers cancelled it. Two months later, the dinner was held. To the organization's good fortune, Lyndon Johnson, Kennedy's successor, agreed to take the slain president's speaking slot at the rescheduled event.

The Weizmann Institute was, and is, a leading Israeli scientific research center founded in 1934 by Chaim Weizmann, a world-renowned scientist who later became Israel's first president. The institute was renamed in Weizmann's honor in 1949, a year after the country was founded, when he was elected the ceremonial head of state. From its earliest days, the institute had taken on an array of scientific challenges. One of these was how to efficiently remove salt from seawater.[1] The desalination research was scientific, but it also had important ideological and political implications for the young country.[2]

Success in desalination would produce important benefits for Israel in helping to fulfill the Zionist goal of building a secure, self-sufficient economy and society that would be a magnet for Jews worldwide. Lacking adequate natural water from rain and rivers, the nation's growing water deficit would be an impediment to both its economic vitality and, as important, its ability to absorb new waves of Jews resettling in Israel. Large-scale desalination of seawater from the Mediterranean was seen as an ideal, if entirely theoretical, solution.

David Ben-Gurion, Israel's first prime minister and the moving force in building the institutions that would lead to the creation of a state, never had water far from his mind. Shimon Peres, Ben-Gurion's close aide and himself later Israel's prime minister and president, says Ben-Gurion talked about water all the time. Ben-Gurion, Peres says, was captivated by the idea of turning salty seawater into freshwater for homes and farms.[3]

Lyndon Johnson shared Ben-Gurion's deep interest in "desalting" water. Coming from a hardscrabble Texas life, Johnson's views about water were similar to the desert-centered Ben-Gurion. A few days before his election in 1960 as Kennedy's vice president, Johnson took time out from campaigning to help prepare a lengthy article for *The New York Times'* Sunday magazine. The article advocated a national focus on developing cost-effective desalination techniques as a tool for eradicating poverty and promoting world peace. Candidates in the heat of a campaign put out many proposals, but Johnson could have placed an article in the magazine on any of several higher profile topics. But he chose to write about what he called "desalted water," a seemingly odd topic for water-rich New Yorkers at any time, and especially so in the closing days of a tight presidential race.[4]

Desalination has the feel of science, engineering, and alchemy combined. The medieval alchemist tried to take lead, a product of scant value, and transform it into one of great worth, gold. So, too, the desalination process tries to take seawater (or inland, brackish water), strip it of its worthless elements, and change it into a lifesaving product of enormous value.

The ancient Romans tried to purify seawater for their army, but their efforts never went far.[5] During World War II, American scientists also began thinking about ways to either take the salt out of the water or the water out of the salt, which sounds like the same thing, but which require completely different approaches and scientific techniques. The problem with either approach, they realized, was that it might make sense in limited military applications where expense is of little concern, but the enormous amount of energy needed to produce pure water from seawater would have made it impossibly expensive for civilian use, at least with then current technology.[6]

Expensive or not, Johnson was sure desalination was in America's and the world's future. He had been instrumental as the Senate majority leader in obtaining funding for federal research on the issue, most of which was allocated to the US Office of Saline Water, which had been established in 1952.[7] Senators knew that Johnson could be counted on to support bills which included water components. And all the more so, when desalination research was included.[8]

"Johnson the Jew"

When Johnson stepped to the podium at the Waldorf-Astoria Hotel ballroom to greet the seventeen hundred dinner guests and Weizmann Institute donors in February 1964, few likely expected Johnson to set in motion a project that on the one hand would spark an immediate firestorm in the Arab world, but on the other would promise a significant boost to Israel's own desalination efforts. Johnson said, "We, like Israel, need to find cheap ways of converting saltwater to freshwater, so let us work together. This nation has begun discussions with representatives of Israel on cooperative research in using nuclear energy to turn salt-water into freshwater. This poses a challenge to our scientific and technical skills. . . . But the opportunities are so vast and the stakes so high that it is worth all of our efforts and worth all of our energy, for water means life, and water means opportunity, and water means prosperity for those who never knew the meaning of those words. Water can banish hunger and can reclaim the desert and change the course of history."[9]

From Damascus to Beirut to Cairo, Johnson's speech was met with fury. One Lebanese newspaper columnist addressed the Texas-born Disciples of Christ church-president as "Johnson the Jew" and said that the speech went "beyond recognition of the birth of Israel to recognition of Israel's future." The Syrian government newspaper called the speech "the ultimate in American support for Israel."[10] Israel's adversaries understood what a secure water future would mean to their sworn enemy.

Although Johnson saw desalination as an essential tool in transforming the Middle East, he may have decided to reach out to Israel due to his respect for Israeli science and the country's rapid and remarkable achievements. With uncanny intuition, Johnson saw in Israel a worthy, if junior, partner who might provide an alternative route to his long-standing dream of desalted water.

Four months after Johnson's Weizmann Institute speech, in June 1964, Israeli Prime Minister Levi Eshkol arrived in Washington, DC, on the first official visit to the US by an Israeli national leader. The US National Security Council had prepared an eleven-point agenda in order of priority to guide Johnson's discussions in the two days of talks. "Joint desalination efforts" was third on the list, and it was the only element of US-Israel cooperation that Johnson would mention in his welcoming comments.[11]

For Eshkol, water had been a central part of his rise in the Zionist bureaucracy and politics. He was one of the founders of Mekorot, Israel's national water utility, and served as its head from its pre-State origins in the 1930s. If Ben-Gurion was the water visionary, Eshkol was the loyal deputy whom Ben-Gurion chose to implement the dream.[12] For personal and strategic reasons, Eshkol was excited to speak with his American counterpart about water, in general, and desalination, in particular.

"It Was Like Going to a Casino"

Nathan Berkman isn't sure that Ben-Gurion or Eshkol ever really believed that freshwater could be extracted from seawater, at least at a cost consumers could bear. Berkman came to Israel from Germany in 1931 as a toddler and has lived his whole life in Israel, except for two years spent in the US earning an MBA at New York University. On his return to Israel in 1960, he needed a job, and he went to work at a fledgling government agency focused on desalination that didn't yet have a name but that would come to be called Israel Desalination Engineering. From his perch in the small government department and with an iconoclastic, unsentimental personality, Berkman was perfectly placed to see desalination grow from a mere concept with different theoretical approaches into a global reality.

Decades later, when Israel's success in desalination was already an established fact, Berkman looked back candidly and remembered the country's first days developing desalinated water. "The beginning of desalination in Israel was Ben-Gurion with a dream," says Berkman, "but, in fact, I don't think that Ben-Gurion actually believed his own dream of desalination was possible." Berkman says that no one in the government agency tasked with exploring desalination actually believed in it.[13]

"The attitude was one of 'You never know'," Berkman continues. "It was like going to a casino, putting down ten chips and hoping something would work. In the early years of the country, we put down a lot of bets on water. The National Water Carrier was one. Rain-cloud seeding was another. There were many. No one really could believe that desalination would work, at least not at any economical price. But we gave it a try."

Berkman also suggested other motivations for Israel's leaders. "For Ben-Gurion and Eshkol," he said, "there was no downside. If it ended up working, the country would have desalinated water. If not, they could go around making inspiring speeches about the dream of making the desert bloom."[14]

Berkman's mild cynicism about Ben-Gurion's motivation and his dream is belied by Ben-Gurion's diaries, which contain many earnest references to the scientific prospects and societal implications of desalination.[15] He seems like a true believer. Ben-Gurion also embraced the ideas of an intriguing "mad scientist" who proffered what Ben-Gurion and many others hoped would be the breakthrough technical solution to purify seawater at nearly no cost.

That "mad scientist" was Alexander Zarchin, who arrived in Israel from the Soviet Union in 1947, shortly before the founding of the state. Trained as a chemist, Zarchin came up with an idea in the early 1930s that he thought would have broad application in water-poor parts of the far-flung Soviet Union: He wanted to desalinate brackish water by freezing it. But before Zarchin could test his hypothesis, his scientific ambitions were interrupted. When the Soviets discovered that the Ukrainian-born Jew from a religious background was a Zionist—a crime in the Soviet Union—Zarchin was sentenced to five years at hard labor in an asphalt mine west of the Ural Mountains. Soon after his release, World War II began, and he was recruited into the Red Army. After the war, Zarchin succeeded in leaving Russia for what would soon be the State of Israel.[16]

Eager to make a contribution in his adopted homeland and quite sure that his freezing technique would solve Israel's water problems, Zarchin, a new immigrant, managed to get an appointment with one government official after another. By 1954, Zarchin, described as a "nudnik" [a pest] in a contemporary news account, finally got face to face with Ben-Gurion.[17] The prime minister wasn't a scientist, but he was intrigued.[18]

Zarchin's idea was built on the scientific principle that when seawater freezes, the salt gets pushed out of the water. Despite the technical challenge of it, if the salt could be rinsed off the frozen water crystals, what remains is salt-free water in frozen form, or ice. Having no value, the extracted salt taken from the water would be disposed of, by recycling it back to the sea, a relatively simple mechanical process. Once the desalted ice was melted, freshwater remained. Zarchin had the further insight—one of enduring significance—that freezing the seawater would best be achieved by spraying it into a vacuum. The vacuum-freezing vapor-compression approach became known as the Zarchin method.[19]

After Israeli scientists reviewed Zarchin's idea, they couldn't tell if it was a transformational breakthrough without building an expensive pilot plant. Ben-Gurion decided to continue funding the concept, a big decision for a budget-constrained country. Ben-Gurion revealed his ambivalence in his diary: "Perhaps Zarchin's invention isn't practical, but it is also possible that it will succeed and we will have a revolution of redemptive force and with international value."[20]

Israel Desalination Engineering was created by the Israeli government to develop Zarchin's idea. Nathan Berkman was one of its first employees. To help defray the costs, and share the risk of building desalination plants to test the Zarchin method, the Israeli government entered into a joint venture with Fairbanks Whitney, a Chicago-based water and industrial company.[21] After years of planning, construction, and testing, it became apparent that this process, which was supposed to be cheaper than extracting well water, was impossibly expensive, more than five cents a gallon versus the thousandth of a penny that Zarchin had projected.

Five cents for a gallon of water may seem inexpensive in a world where bottled water costs more per gallon than gasoline. But for households, and especially for agricultural use, five cents a gallon would quickly make food too expensive to grow. Despite these early setbacks, Ben-Gurion, Eshkol, and their US partner decided to push on for several more years, with Zarchin blaming the failure of his system on the inability of others around him to understand his invention.[22]

Finally, Zarchin left the government agency to pursue other ideas, and went from one project to the next, complaining of being misunderstood at each one.[23] Yet, he left behind a significant legacy.

The Israeli government went from talking about desalination to actually doing something about it. In this still fledgling field, Israel now had a highly experienced cadre of trained professionals. Further, Zarchin's use of evaporation and vapor compression would be utilized again a few years later as part of a different method, but this time with great success. The technical expertise that originated with Zarchin would prove to be a game changer for Israel.[24]

Perhaps most important, following Zarchin and Israel's early jump into desalination, President Johnson, a close student of desalination efforts, took note that Israel shared his interest in "desalted water."

Distracted by War

Eshkol's 1964 visit to Washington, DC, came at a dynamic time for Israel. The massive, and massively expensive, National Water Carrier was due to open a few days later. The country's economy had begun to take off and its population was growing at a rapid pace.[25] The combination of more people and more business had put a strain on Israel's water resources, while the nation-building activities and security expenditures to protect against episodic terror attacks drained its budget. Eshkol and his diplomatic party were delighted to use the prestige of a state visit to deepen the US-Israel relationship, which was not nearly as close and interwoven as it is today. But the trip was not purely symbolic for the prime minister. Eshkol hoped the meeting would produce significant US financial support to accelerate the development of Israeli desalination.

Johnson was known for being tough or sweet depending on what he wanted. With Eshkol, he was all charm. At the state dinner on the first evening of the two-day visit, Johnson toasted his guest, saying, "Mr. Prime Minister, you told me only this morning that water was blood for Israel.

So we shall make a joint attack on Israel's water shortage through the highly promising technique of desalting. Indeed, let us hope that this technique will bring benefit to all of the peoples of the parched Middle East."[26] For Eshkol, this was a good start, but not what he was after.

The following day, the two men met again. As with his toast, Johnson focused his attention on desalination. He told Eshkol, "We want Israel to have more water. For that reason, we are ready to undertake a study in connection with the desalting program which will also provide Israel with the water it needs. If the study proves that the desalting project is feasible, we will help in working it out."[27]

The two great expenses in desalination are in the ongoing energy costs to desalinate the water and the one-time building costs to create the facility. A country like Saudi Arabia with limitless energy supplies and a budget surplus for funding capital projects could, and did, have a large desalination plant, even if it was expensive and energy inefficient. Israel then had virtually no known energy resources and would have to buy expensive, imported coal, oil, or gas to keep the plant running.

Regardless of energy source, once the amortized cost of building a desalination plant was added to the cost of the imported energy to run it, Eshkol knew it was out of his country's price range. Even with Israel's then recent growth and success, especially with an outcome that was unclear, Israel's budget couldn't shoulder the tens of millions or more in expenses to build an industrial-scale facility. For that, he would need American help.

Johnson wanted Eshkol to understand the geopolitical price he was prepared to pay to help Israel get started. "We will help Israel on this [desalination program] as much as possible," the diplomatic minutes of the meeting show Johnson as saying. "Of course, we will get some backlash from the Arab countries as a result of your visit with me. However, I am not worried by that. It is important both to you and to the United States that everybody should know we are friends. So, that is no reason to not go forward with this desalting project."[28]

Eshkol sought just one more piece, and that was word of financial support for Israeli desalination. Although there is no record of Johnson explicitly making an offer in the diplomatic minutes, the pre-meeting memo prepared for the president says that American loans to Israel for a desalination plant could go as high as $100 million, a huge sum for that era.[29] Although a loan and not a grant, the amount discussed in the memo was more than double what the US government had spent—in the aggregate—on the work of the Office of Saline Water from its creation in 1952 through the early 1960s.[30]

Closing the two days of talks in Washington, DC, the president went with the prime minister to a women's reception at the Mayflower Hotel hosted by Eshkol's wife, Miriam, for wives of official Washington. As seemed to be the protocol of the day, the event was covered by female reporters. With a group of newspaperwomen encircling them, the two leaders chatted warmly, almost playfully. Johnson told the journalists, "We are very much alike, we are both farmers." He then told the press corps of an exciting new program Israel and the US would be developing in desalination that would "remake the world without deserts or droughts."[31]

Wise to the ways of bureaucracies and determined to not let his vision fall prey to plans other than his, Johnson set up negotiating teams to work out desalination details with the Israelis and established the Interagency Committee on Foreign Desalting Programs to figure out financing for the Israeli initiative.[32] But by early 1966, Johnson's Secretary of State, Dean Rusk, told him that the Israeli desalination plant was delayed because of "a host of political, economic, and financial considerations."[33] Johnson decided to appoint Ellsworth Bunker,[34] a prominent diplomat, as his special envoy, and sent him to Israel that December. While there, Bunker proposed to Eshkol the construction of a $200 million desalination plant, but Bunker also wanted to know how Eshkol thought costs should be split and the US loans repaid.[35]

As significant as desalination clearly was to Johnson, by late 1966, the Vietnam War was consuming much of his time and focus. Bunker was reassigned by Johnson as ambassador to Vietnam and no replacement was selected to coordinate desalination ideas with the Israeli government. For Eshkol's part, water wasn't at the top of his agenda either. The winds of war in the Middle East were beginning to blow. As Johnson was preoccupied with Vietnam, Eshkol was distracted by what grew into the Six-Day War in June 1967.

The next time Eshkol saw Johnson, in January 1968, it was as a warmly welcomed guest at the LBJ Ranch, Johnson's home in Texas. On this visit, desalination was low on the list of Eshkol's and Johnson's interests or concerns.[36] In the buildup to the Six-Day War, Israel's long-time weapons' supplier, France, decided to switch sides and befriend the Arab states. While projecting an air of confidence from the smashing victories over its Arab adversaries, Eshkol and his military advisors knew that the Arab states would rebuild and rearm. Israel needed to find its own source of weapons. Eshkol hoped the US would be that source.

Although Johnson disappointed his guest by deferring a decision on weapons' sales to Israel, he offered Eshkol a kind of consolation prize. He ended the three-day visit with a promise to Eshkol that the US would reengage on bringing desalinated water to Israel.[37] Johnson soon appointed George Woods, a former World Bank president, to take over talks on desalination with Israel. Woods consulted with the Israelis, and then met with Johnson soon after the November presidential elections in which Johnson had chosen not to run.

In their meeting, Woods proposed to Johnson that a $40 million grant and an $18 million loan be given to Israel. After being assured that this funding would be sufficient to build a small desalination demonstration plant, if not a mega-facility, Johnson said that he was ready to move forward.[38]

Three days before leaving office, with all of the activities and obligations of the moment, Johnson sent a letter to Eshkol in which he recounted his personal efforts to help Israel develop desalinated water. He wrote Eshkol that, as one of his last official acts, he had made a request that Congress provide full funding for the Israeli desalination plant. He mentioned with a mixture of pride and excitement that the facility should soon be producing "forty million gallons a day of desalted water." Eshkol replied immediately that Johnson's actions were crucial for the advancement of economic progress and the establishment of peace in the Middle East.[39]

The Thirteen Engineers Who Changed the World

Johnson's promised investment in an Israeli desalination plant was delayed repeatedly. Incoming President Richard Nixon, who had supported funding the Israeli project during the presidential campaign, had his own plans for Johnson's budgeted, but not disbursed, money when he came to office.[40] And even more pertinent, the process of handing out government money can grind slowly, especially when those writing the checks—here, the US Office of Saline Water—and those who deposit them—an eager Israeli government—had different visions of how that money should be spent. Nathan Berkman had no illusions of the US government sending money without a breakthrough idea from Israel. The Americans, he remembers thinking, had enough projects of their own that needed funding.[41]

Israel's entire desalination research apparatus—the equivalent in national significance, if not size, to the US Office of Saline Water—was a small bureau set up in 1959 to bring Zarchin's ideas to life. When Zarchin's idea proved to be unrealizable, and he left the bureau in 1966, Nathan Berkman—six years after receiving his MBA in New York and starting his first job—took over the group.

"When you see Israel's role in desalination today, it is hard to understand what it was like in the early days," says Berkman. "After Zarchin's idea failed, we knew we'd need another technology. This was as much about us keeping our jobs as it was searching for a great breakthrough."[42]

At the time, the Israel Desalination Engineering department consisted of all of thirteen engineers. Once a week, Berkman—who had studied economics and business administration and who was not an engineer—would convene a meeting of the engineers, primarily to brainstorm new approaches to desalination but also to evaluate ideas being tried elsewhere in the world.

Berkman's team of engineers began developing some of the ideas that grew out of the brainstorming sessions, and this at a time when there was still no settled approach to efficiently and economically desalinating seawater. By combining certain mechanical elements of Zarchin's technique unrelated to the freezing and coupling them with various concepts for heating water to create vapor, Berkman's team created two new energy-efficient approaches to desalination, both of which remain in use around the world today.

The first idea, called Mechanical Vapor Compression, or MVC, is very reliable and used in settings where the cost of an unscheduled shutdown would be economically unacceptable. MVC is utilized in mining operations where a failure to have freshwater at hand would mean that the water-intensive mining would have to come to a halt. The negative aspect of MVC is that the assurance of consistency comes with a price in higher operating costs, thereby making it less desirable for large-scale production of desalinated water.

Research around MVC, though, led to a second approach invented by the Israel Desalination Engineering team, which was a variation on a heat-based process called MED, short for Multi-Effect Distillation. MED had been invented in the late 1800s and was used in a variety of industrial processes such as evaporating fruit juice to obtain its natural sugar. Similarly, MED began to be used for extracting freshwater from seawater.

The Israel Desalination Engineering team's MED innovation was to use a series of linked aluminum tubes to replace the chambers traditionally employed to heat the water to produce vapor. Since these aluminum tubes held and transferred heat more efficiently than any previous method or material, the temperature could be kept consistently high, thereby reducing the need for a new energy source to heat water added during the process. Israeli MED used less energy than any other heat-based desalination process. In a heat-intensive process, it was a giant leap forward in reducing costs in the then still evolving field of desalination.[43] But when first introduced in the late 1960s, Israeli MED was still viewed by many as an interesting theory with no certainty that it would work when installed in a real-world setting.

Skepticism about aluminum-tube MED would only change after a demonstration plant was built in the mid-1980s. With $20 million authorized as a special US grant and matched with Israeli funds,[44] a demonstration plant was built in Ashdod, on Israel's southern Mediterranean coast.[45] The American gift was the fruit of a long process initiated by President Johnson, even if modified several times by others in the intervening years.

Despite Israeli frustration that the money didn't arrive sooner, the methodical, bureaucratic pace set by the Office of Saline Water served Israel well.[46]

The Israeli facility was built with less than the $100 million Johnson had considered in 1964. But these grand dreams about desalination in Israel played their part in moving the US government bureaucracy forward. The demonstration plant was too small to enhance Israel's freshwater supplies, but large enough to prove Israeli MED's effectiveness as a next step in desalination's development and implementation.

Starting a Global Company for Self-Preservation

While aluminum-tube MED was still in development in the 1960s, Berkman began to grow concerned about a potential problem far more immediate and personal. With Zarchin's freezing technique out of the picture and with money from the US that wasn't about to be disbursed any time soon, Berkman began to fear that Israel Desalination Engineering was at risk of being eliminated at the next annual budget review.

Keen on keeping his job, Berkman decided that his group had to figure out a way to break free of its reliance on Israel's national treasury.

With other countries looking for desalination solutions to their own water problems, in the late 1960s and early 1970s, Berkman decided to have his government division go into business and to start selling the group's desalination know-how—such as it was—to others. "I did this on my own," says Berkman. "I didn't clear it with anyone first. I wanted to be as invisible as possible so that the next time someone looked at us they would see us costing the government nothing. I also thought it would be a good idea for us to develop practical experience with desalination plants."[47] With this, the government unit became a profit-seeking business embedded within the government.

Berkman changed Israel Desalination Engineering's name to IDE Technologies and announced the group's availability to build desalination units. Shortly after, IDE received its first contract to install an MVC facility in the Canary Islands, a small archipelago that is part of Spain. A contract with the then still friendly Iranian government to install small MVC desalination facilities at several of its air force bases followed. The Israeli government desalination company thrived.

IDE was merged with another Israeli-government business in the 1980s, and as part of the privatization boom in Israel in the 1990s, the company was sold.[48] During Nathan Berkman's twenty-five-year tenure overseeing his increasingly active for-profit government agency, IDE designed, constructed, installed, and/or managed more than three hundred desalination plants. Many of the facilities from the Berkman era were small, but "with each one, we tried something new," says Fredi Lokiec, now a senior executive at IDE. "We never accepted the idea of a carbon-copy approach. With each plant, then and now, we looked to see what we could do differently."[49]

In recent years, IDE has designed and built many of the world's largest desalination plants. The largest desalination facility in the Western Hemisphere is in Carlsbad, California, producing fifty-four million gallons of freshwater a day.[50] The largest desalination plants in China[51] (fifty-three million gallons a day) and India[52] (106 million gallons a day) are also IDE desalination plants. The China facility is exclusively an MED plant and the Indian one is an MED hybrid. Likewise, Mekorot, Israel's national water utility, has developed its own footprint in building or managing desalination plants, even if on a smaller level.

Not surprisingly, IDE has played a central role in Israel's development of its own desalinated water supply. Three of Israel's largest desalination plants were built and are managed by IDE, including the world's largest and most modern one, a facility in Soreq, about ten miles south of Tel Aviv, producing 165 million gallons a day.[53]

The Road to Soreq

Ben-Gurion may have dreamed of a day when the omnipresence of desalinated water would transform the country and the region, but he could deliver inspiring speeches about making the desert bloom without making any mention of cost. And when that cost was measured against free or very cheap water from rain, aquifers, or the Sea of Galilee, a chorus of opposition consistently drowned out the dreamers. Although the price of desalinated water had dropped fifteenfold in the decades from its first production in the late 1950s utilizing the Zarchin method, it was still in the early 1990s an expensive source of water.

Ronen Wolfman was a senior official in the Ministry of Finance's Budget Department in charge of infrastructure in the 1990s. He counted himself among the fierce opponents of building Israel's seawater desalination capacity at that time. "First," says Wolfman, "the price was too high. The technology was getting better all the time and I didn't want to commit to some huge plant that we'd have to live with for decades. I wanted to push the day out as far as possible. Second, we still weren't using enough of our treated sewage for agriculture. I didn't want to do anything that would discourage that. And third, I was pretty sure our farmers were picking the wrong crops because we had been too lenient with our water allocations to them. With a change in crops, we could save a lot of water."[54]

The Budget Department worked with other government agencies to get farmers to plant more water-efficient crops. The amount of cotton—a water-intensive crop—planted in Israel was cut by about seventy percent. Scientific grants were given for research into plants that would grow without large amounts of water. And the government encouraged an all-out push to capture as much of the country's wastewater as possible in order to reuse it for agriculture.[55]

Prayers for rain may also have been heard and heeded from on high—but with the effect of slowing the move to desalination. Every few years, Israel enjoyed a winter of heavy rainfall. Each time the Sea of Galilee and the country's aquifers became refilled, policy makers at the Israel Water Commission and inside the Ministry of Finance would use it as a reason to defer any commitment on desalination.

Two professors at the Technion thought the indecision was irresponsible and wanted to try to reorient national water priorities. "This was before anyone was talking about climate change, but we knew it was just a question of time before Israel would have another drought," says Professor Rafi Semiat, an expert on desalination. "Inevitably, one of these droughts would overwhelm us."[56]

Semiat and a Technion colleague, Professor David Hasson, decided to try influencing the policy-making process. "Aware of the long lead times for infrastructure projects," Semiat says, "we wanted our leaders to stop using a rainy year or two as an excuse to not make the inevitable decision that we needed an advanced technological solution to this problem."[57] The professors created the Israel Desalination Society and convened an annual meeting beginning in 1995 to educate and to advocate for desalination as a solution.

Even more important, leading figures in both of the dominant political parties came to believe that gambling on hardy winter rains and relatively easy fixes would not be enough.[58] Future prime minister but then minister of national infrastructure, Ariel Sharon concluded that sooner rather than later, desalination would become a necessity.[59] When the government that Sharon was serving in was replaced in the 1999 national elections, the new minister of finance, Avraham Baiga Shochat, had independently come to the conclusion that desalination had to be studied closely.

Shochat had earlier served a term as minister of finance beginning in 1992, and in that capacity he had engaged in discussions about an infrastructure fix to the water problem, anticipating a time when a severe drought would hit the region. But during his first tenure as minister of finance, consideration of desalination and other solutions were tabled whenever heavy rains came. In late 1999, back in power and again with the Finance portfolio, Shochat convened a group of Cabinet ministers to help create a consensus on whether the time was right for the jump into desalination.[60]

Before moving forward, Shochat wanted to be sure there was no other alternative that would work. One proposal that received a lot of attention was importing water from Turkey,[61] then a close political and military ally of Israel. Although Turkey today has water problems largely tied to mismanagement and overuse, at that time, in 2001, it was still a water-rich nation that wanted to monetize its water surplus. Israel's military was also excited by the idea as it would deepen the connection between the two countries.[62] But after reviewing many approaches, the Turkish option died because of price considerations as well as logistical ones. Considering the Turkish antagonism for Israel today, the decision to not be reliant on Turkey for Israel's water proved also to be wise geopolitically.[63]

Another key decision Shochat and the committee of Cabinet ministers had to make was whether the government should build the desalination facility or whether it should be put out for bids and run by a private company under strict contractual parameters. Israel had a long history of using government entities for major projects, and Mekorot, the all-purpose national water utility, had a depth of experience in desalinating inland brackish water.

But Ronen Wolfman and others in the Budget Department had recently developed a highway using a private company. Although Wolfman was not a supporter of building desalination facilities, he was sure that a private entity would be superior to Mekorot or any other government body, even with Mekorot's desalination experience. "Not only would the cost of the project get placed on someone else's shoulders, but I was sure the private sector would come up with a better outcome than the government or even better than a government-owned company like Mekorot," says the kibbutz-born and -raised Wolfman who, ironically, later became Mekorot's CEO and who is now one of the leaders of Hutchison Water, a Chinese-Israeli water company that is part owner of a large Israeli desalination facility.[64]

The Cabinet chose to go the private route, picking a consortium of Israel's IDE, France's Veolia, and others. A key consideration in selecting this bid was not only the group's expertise in desalination, and its financing capability, but the certainty that the partners had the management depth to operate the desalination plant for the twenty-five-year life of the agreement. The plant would be built on Israel's Mediterranean coast in Ashkelon, operated by IDE-Veolia, and, at the end of the term, the group would transfer ownership back to the government. In return, the government committed to buying a fixed amount of water each year along with paying a defined annual sum so that the partners would have a guaranteed cash flow.[65]

In designing the new seawater desalination plant, IDE and Veolia had several important decisions to make, perhaps the most important of which was deciding which technology to use to desalinate the salt water. Years earlier, IDE had changed the world of desalination with its aluminum-tube MED approach and had successfully utilized it in dozens of plants built in other countries. It would be logical to use MED in Ashkelon, too. Instead, the partners decided to make use of an even more energy-efficient idea called reverse osmosis, or RO, with which IDE had far less experience. Coincidentally, reverse osmosis had a strong Israel connection.

Reverse Osmosis: A Giant Step Forward

Seawater is a mixture of pure water, salt, and other minerals. When seawater goes through reverse osmosis, the water is pushed through a membrane with the pure water sent in one direction and the salt molecules in another. The segregated salty slush that remains is called brine and is returned to the sea. The same process can be used to take minerals or other unwanted material out of the source water. But whatever the particles to be removed are, the essential element is the membrane.

The RO membrane wasn't originally created with seawater in mind, but rather for brackish water. Less salty than seawater, brackish water is found in so-called fossil aquifers that hold water collected in prior geologic eras and left undisturbed long enough for greater or lesser amounts of salt and minerals to leach into the underground water source. Brackish water is also created when seawater and freshwater come into contact with each other, as when a river meets the sea.

In his mid-forties, the Kansas-born Sidney Loeb began pursuing a Ph.D. at UCLA in the early 1960s in an evolving field of chemical engineering. Loeb was investigating whether brackish water might be purified to drinkable freshwater via a specially constructed membrane. Working with a lab partner, Loeb developed a membrane that had nano-sized holes that were large enough to allow pure water to pass through but small enough to block particles of salt and other dissolved minerals.[66]

In 1965, Loeb's membrane was brought to the small California town of Coalinga for a test. The town's available water was so densely laden with minerals that it was undrinkable. Whatever water the town needed had to be delivered by train from another town. The test not only changed the future of Coalinga, but also of desalination—as Loeb's membrane successfully purified the previously non-potable water.[67]

Had Loeb had the instincts of a businessman, he'd have marketed his RO membrane. Instead, he filed a patent and then did nothing. Around that same time, his marriage began to fall apart. Because Loeb needed a job and because California permitted divorce without litigation if the couple were apart for a year, in 1966, he accepted an opportunity to come to Israel for a nine-month project. It was his first visit to the country. Loeb was reborn by his visit. The nine months turned into a lifetime.[68]

A scientific nonentity who no one listened to back home, Loeb was embraced in Israel as a deep thinker. The Negev desert was believed to have trillions of gallons of mineral-laden water beneath its sands. Some of the fossil aquifers there were used for the region's meager development, but the inland desalination effort was expensive and often thought not worth the trouble. Loeb's membrane was seen as a possible solution. An experiment desalinating brackish water in the southern Negev at Kibbutz Yotvata was a great success. Loeb became a local sensation in the scientific community.[69]

Beyond professional recognition, Loeb fell in love with a woman who had moved to Israel from England in 1946 while still a teenager. And although he was largely an assimilated American Jew, Loeb's visit to Israel coincided with the Six-Day War, which awakened in him deep feelings of national identity.

Loeb stayed in Israel for nearly three years, and then returned to Los Angeles to divorce his wife. He soon married Mickey, his British-Israeli love. But the esteem he received in Israel wasn't transferable back to the US. There were no academic positions offered to him upon his stateside return and a brief effort to set up a consulting company dissolved almost as soon as it started. When he received a call from the head of what was to become Ben-Gurion University (then the Negev Institute) inviting him to join the faculty in chemical engineering, he jumped at the chance.[70]

Back in Israel, Loeb presented his RO membrane to Nathan Berkman and others, but they didn't believe it was a better solution than the IDE's MED process.[71] "Sid had no business skills and didn't know how to sell himself or his services," says Mickey. "In the lifetime of his reverse osmosis patent, he received fourteen thousand dollars in royalties for it. Can you imagine that? Fourteen thousand dollars for an idea that launched a multi-billion-dollar industry!"[72]

A more impartial observer agrees with Mickey Loeb's assessment. "Sidney Loeb was to reverse osmosis what the Wright Brothers were to aviation, Henry Ford to the car, and Thomas Edison to the lightbulb," says Tom Pankratz, an American, an industry veteran, and the editor of the *Water Desalination Report*. "For sure, others have made all of these better after they started it, but they were the founders. Sidney Loeb was the father of reverse osmosis. He just didn't get the rewards in fame or money that the others received."[73]

By all accounts, Loeb was a man of exceptional kindness and modesty. But living to see his RO membrane adopted for the Ashkelon desalination plant was a validation of his life's work. He attended the opening in 2005, but died three years later before he could see how seawater reverse-osmosis desalination would change Israel and the world. From a limited usage in removing salt and minerals from inland brackish water, reverse osmosis today is the dominant technology, already responsible for purifying sixty percent of the salty water desalinated at facilities around the world.[74] As older desalination plants are retired, the role of membrane filtration will only grow.

"A Period Like the Dawn of Agriculture"

Desalinated water is manufactured, and will always be more expensive than natural sources like rain, lakes, rivers, or even aquifers. But the Ashkelon plant offered a surprise. Due to the use of the RO membrane, the water wasn't just the highest quality water to be found in Israel in terms of cleanliness, low salinity, and high clarity; it also turned out to be about fifty percent cheaper than any of the cost estimates the Cabinet had received when deciding to pursue desalinated water. At such a low price, the government then asked IDE-Veolia to double the Ashkelon plant's daily production to eighty million gallons of desalinated water.[75]

But Ashkelon, when it opened in late 2005, was just the start.

Seawater desalination plants opened along Israel's Mediterranean coast in Palmachim in 2007 and then in Hadera in 2009. The desalination mega-plant Soreq opened in 2013. While both Palmachim and Hadera introduced significant innovations of their own, size aside, Soreq is a marvel of engineering and financing creativity. To drive down costs, it makes use of spurts of lower-cost or off-peak electricity throughout the day and night, a feat that may sound easy but isn't.[76]

Because beachfront property is so valuable, Soreq was built about a mile from the Mediterranean and about two miles from where it both takes in seawater and then, by a separate pipe, returns its hyper-salty brine to the sea via a diffuser. Due to the presence of built-up areas between Soreq and the seashore, the large pipes had to be installed under the ground largely without the benefit of excavation. The channels to the Mediterranean were constructed by pipe jacking, a process of ramming pipe components with great force to cut through the earth below.

At a $400 million construction cost, Soreq may not be emulated everywhere, but its pathbreaking technology, especially in energy management and savings, will trickle down to all RO plants that will follow. Built by IDE and the Chinese-Israeli Hutchison Water, the plant incorporates environmental protections for clean water and fish safety that were barely contemplated only a few years earlier.

With Ashkelon, Palmachim, Hadera, Soreq, and a Mekorot-managed plant in Ashdod, along with RO desalination plants for brackish water, Israel now produces nearly five hundred million gallons of freshwater from salty sources every day. Ten years earlier, there had been a few brackish water desalination devices at work, and a small seawater desalination facility in Eilat, Israel's southernmost city, far from the Mediterranean. From nearly zero a decade ago, if desalinated water could be piped exclusively to homes and not part of the freshwater mix including aquifers, wells, and the Sea of Galilee, it would now be the equivalent of ninety-four percent of Israel's household water.[77]

Desalination has entirely transformed the water profile of Israel, and its effects have been felt throughout the entire society. The extensive use of desalinated water has had enormous implications for Israel's environment, economy, infrastructure, social harmony, public health, and even its relationship with its Palestinian, Jordanian, and other neighbors. In each of these realms, Israel has begun to see benefits that are likely to grow over time.

Most obvious, having a lot of new water changes Israel's relationship with nature and recent changes in weather patterns. "We have gotten ahead of the climate change question," says Shimon Tal, a former head of the Israel Water Commission. "It isn't *just* desalination, but with all of this new desalinated water added to everything else we are doing, we are mostly immune from adverse weather conditions. Droughts have plagued the Middle East since the time of the Bible. Israel now can withstand even a long drought. Because of this, farmers and businesses can plan without unwanted disruptions from nature."[78] Israel's consistently strong economic performance will still be subject to business cycles and global competition, but a shortage of water will not stymie growth in industry, tourism, or agriculture.

As tempting as it is to think that desalination by itself solved all of Israel's water problems, that isn't entirely the case. It is the combination of many different approaches and techniques that gives Israel its water security. Desalination may be the most valuable part of the mix, but it cannot succeed standing alone. It is too expensive, and the security risk too great, to allow it to become the only or even majority source of Israel's water.

But, all of this new desalinated water does address one of Israel's many national security concerns. With agriculture being such a small contributor to GDP and with foreign currency reserves in place, Israel could have decided to phase out its agricultural sector, utilizing imported food over homegrown supplies. With that one move alone, there would have been no need to build desalination plants or to ask its citizens to limit daily use. But, Israel's strategic planners are ever mindful of Israel's regional isolation and the instability of its geopolitical situation. While Israel doesn't grow all of its own food—it imports much of its animal feed stocks—the country seeks self-sufficiency, or adequacy, in homegrown food. With its water supply assured, it is unlikely that any combination of war, embargo, or drought would lead to Israel finding itself without food for its people.

"Desalination means that you don't have to rely on others," says Ilan Cohen, the director-general of the Prime Minister's Office under Ariel Sharon and Ehud Olmert. "Even when we were cutting other parts of the national budget, we moved forward with infrastructure for desalination. It lets us control our destiny, something important for any country, but especially for us for as long as we are surrounded by enemies."[79]

Moreover, agriculture in Israel, as with most other countries, is more than just a source of food or a contributor to the economy. It plays an important social role that belies its two-and-a-half percent contribution to GDP. Israel's population is densely concentrated in a relative handful of urban bands. Since cities have few open green spaces and since no one is more than a short distance to farms and fields in the small country, agricultural areas serve as a key part of the country's natural landscape.

Farms not only broaden the national footprint and help to limit urban sprawl, but also continue the tradition of placing communities adjacent to the country's borders.[80] This began as a security buffer against infiltration, but also sets the borders in a region filled with parties who often test boundaries.

All of the new water will serve as an accelerating benefit to the environment in Israel. Already, there is more water flowing through Israel's rivers, and the new water helps address what was becoming an urgent concern: Israel's aquifers were increasingly at risk of overpumping.

In operating the five Mediterranean desalination plants, Israel could reduce costs by not removing all of the salt. It would not have been illogical to bring the saline level down to match the salt profile of the Sea of Galilee and Israel's other water sources. Since humans don't taste salt until it reaches highly elevated levels, there was no requirement to reduce it to the ultra-low salt levels that were achieved. But as a result of this desalting—combined with the fact that Israel blends all of its water sources into an integrated water supply—the introduction of highly desalinated, essentially salt-free water means that the total salt content ingested by the nation will decline, with positive implications for agriculture and public health.

Water used for irrigation will be less salty, putting less strain on soil and crops. Aside from possible health benefits from lowering the salt content of everyone's drinking water, the desalination process will also lower the concentration of nitrates found in the Coastal Aquifer as it will have a chance to recharge and be diluted, to the benefit of pregnant women and their babies.[81]

All of these positive effects—and others like a slower replacement rate for the nation's boilers and industrial machines due to less mineral buildup—has had an unexpected economic impact contributing several hundred million dollars a year to GDP and effectively lowering the cost of the desalinated water by about a third from its already lower-than-expected price.[82] With Israel's recent discovery of enormous fields of natural gas just offshore, which can be used to power the desalination plants, the total costs will continue to fall as the benefits continue to accrue.

One intangible benefit of Israel's growing supply of new water is opportunities for deepened cooperation with its neighbors. Under its 1994 peace treaty with the Kingdom of Jordan and the 1995 Oslo II Agreement with the Palestinian Authority, Israel provides water to each.[83] To the extent that climate change, population growth, or prosperity create new water needs for the Palestinians or Jordan, Israel's ability to produce more desalinated water puts Israel in the position of being able to backstop its neighbors until weather patterns change or they create adequate alternative water sources of their own. The interdependence of the parties will create new opportunities for coexistence and possibly even serve as a prelude to warmer relations.

For Israel, too, as its population grows and as the natural supply of water shrinks, Israel now has the confidence that regardless of population growth or the water needs of its economy, Israelis will not have to worry about water. Although there are no additional desalination plants now under construction, planners have reserved locations for more of them, if needed.

From its origins, Israel has had to build its society without natural resources like abundant water or energy in the form of oil or gas. This led to the elevation of brainpower and innovation as the key drivers in Israel's economy and as the key vehicle for leapfrogging out of its region to the larger world. With the recently discovered natural gas off Israel's coast and the potentially significant shale reserves in Israel's Negev desert that have yet to be developed at a commercially reasonable price, Israel's economic model may be changing in the coming years. If it does, the abundance of water—in significant part assured by desalination—will assist and speed that process. Israel is likely to long continue to be the "Start-Up Nation," but it may also mature into a "Resource Nation."

Israel's water companies will help the country to be a global leader in desalination, providing a further boost to a science-based economy. From new membranes to use of off-peak electricity algorithms to large desalination plant construction itself, Israel is among a very few nations leading the way in all of these desalination components.

The global water crisis is unlikely to be solved without widespread use of desalinated water. Today, in an accelerating migration trend, nearly half of the world's population lives a relatively short water pipeline away from a seashore.[84] Whether for agriculture, industry, or household use, countries and regions will increasingly have no choice but to supplement existing water supplies. Even water-rich locations like New York City may decide to build a desalination plant as a backup for security or environmental reasons. While helping to solve its own water problems, Israel's experience and know-how in desalination is likely to be called upon by many others.

Desalination, says Ilan Cohen, the former top aide to Prime Ministers Sharon and Olmert, will change the way we think about water. "Water is no longer a resource and it should no longer be thought of in that way. With desalination, water has become a purely economic issue. Water is no longer a question of how, but how much. If you think of water as something to be manufactured, it exclusively becomes a question of cost. You can get the quantity and quality of water you want, provided you are prepared to pay for it."

For Cohen, the revolutionary nature of desalination evokes an earlier revolution. "For us, water today is just like food was in ancient times," he says. "It became a paradigm shift when man could grow his own food. When we began desalination and reusing wastewater, it was a paradigm shift. Today, we are in a period like the dawn of agriculture. Prehistoric man had to go where the food was. Now, agriculture is an industry. Until recently, we had to go where the water was. But, no longer."[85]

In Israel today, desalination is only one element of a flexible, integrated, sophisticated program of water governance, but, over time, it is likely to be seen as the most important of all. Even in the earliest days of desalination, otherwise pragmatic leaders like Lyndon Johnson, David Ben-Gurion, and Levi Eshkol dreamed about how desalination could fight poverty and spur peace around the world. Peace may still be elusive, but desalination is no longer a dream.

Seven

RENEWING THE WATER OF ISRAEL

Rivers don't do anything; things are done to rivers.
—David Pargament

THE MACCABIAH GAMES are sometimes called the Jewish Olympics. Held every four years, they bring the best Jewish athletes in the world together for two weeks. For many Jews, participation in the Maccabiah Games, whether as an athlete or a spectator, is their most important lifelong connection to Israel.

Twenty years from conception to the first games in 1932, the Maccabiah was planned as a global competition that would take place in the Land of Israel, contrasting it with the Olympics on which it was modeled, which changes venues with each new Olympiad. The first Maccabiah Games attracted 390 athletes who came from eighteen countries.

There was a second Maccabiah in 1935, but the rise of Nazism in Europe kept the games from being held again until 1950. Eight hundred athletes from nineteen countries participated in what were the first games to be held in an independent State of Israel. The competition was the first major global Jewish communal gathering after the Holocaust.

By 1997, the Maccabiah Games had become locked into a fixed four-year schedule. It was a hot ticket. A fifty-thousand-spectator stadium had been constructed in a suburb of Tel Aviv for the opening ceremony and key events. The games brought visitors from around the world, and more than five thousand competitors from thirty-six countries.[1] The Maccabiah Games had also become a statement of Israel's centrality to world Jewry. But on the night before the 1997 games began, tragedy struck.

A temporary pedestrian bridge had been erected over the Yarkon River for the opening ceremony. The athletes and their coaches, waving their national flags, were to march from the far bank of the river into the stadium. The teams were staged alphabetically. The Austrian team, the first to step on the bridge, mostly made it across.[2] But when the weight of the nearly four hundred athletes and coaches of the Australian Maccabiah team—the second alphabetically, in Hebrew, after Austria—was added to the bridge, it caused the sixty-foot span to collapse. Many of the Austrians and most of the Australians fell into the river.[3]

Remarkably, with bodies falling on top of bodies in the unlit river, no one drowned that night, and tragic though it was, miraculously, only one participant, a bowler from Sydney, died at the scene of injuries suffered from the fall. A few dozen others were hospitalized with broken bones or from water inhalation, but even if many of the athletes would be out of the games, the feeling must have been that it could have been much worse.[4]

By the next morning, though, the athletes' status had deteriorated. Overnight, seven of the hospitalized Australians had been downgraded to critical condition and within a few weeks, three of the previously fit athletes died. Doctors and researchers quickly discovered that the sediment in the Yarkon River was highly polluted. The collapsed bridge and all of the falling bodies had churned up the riverbed. In their brief time in the water, the athletes had apparently inhaled the noxious brew from the river bottom. An Israeli environmentalist called the Yarkon "a trap of stench, dirt and death."[5]

The bridge collapse was a source of shame and soul-searching in Israel. At its proudest international event—while welcoming world Jewry—Israel failed to provide a safe environment for its visitors. Blame was assigned widely: to the designers of the bridge, to the Maccabiah organizing committee, to Israeli society as a whole. Even Prime Minister Benjamin Netanyahu, then in his first term at the head of government, was taken to task for permitting the opening ceremony to continue after the bridge fell. Michael Oren, who would go on to become ambassador to the United States twelve years later in a subsequent Netanyahu government, told a newspaper reporter at the time that the episode was representative of "the rot at the core of our society."[6]

Popular Israeli reaction was initially focused on the outrage over the shoddy construction and disagreement about whether the opening ceremony should have been cancelled. Several people involved in the design, construction, and oversight of the failed bridge were tried, convicted, and served prison sentences. But the episode raised broad awareness about a problem previously known mostly by officials and environmentalists, that "the Yarkon, the river that flowed in the most heavily populated area of Israel, was in a disgraceful state and [that] this needed to be changed."[7]

Over the following years, all of Israel's rivers underwent renovations, restoration, and repair, though more needs to be done to return them to their optimum condition. While environmental laws and enforcement of regulations helped bring Israel's rivers back to life, what may have helped most was that Israel developed new water sources. This new abundance in water—and the now unceasing demand for sewage to treat and reuse—took pressure off all of Israel's rivers. Less water needs to be extracted, less sewage finds its way to the rivers, and there is more surplus water available to add to enhance the flow where and when needed most.

The repair of Israel's waterways was a long process, beginning around the time of the Maccabiah bridge collapse; and it is still ongoing. But how Israel came to rethink its rivers—especially the role of surplus water in their restoration—is a model for communities and countries around the world.

"People Need Access to Nature"

While environmentalism in the Land of Israel is as old as the Bible[8] and the Zionist pioneers celebrated their return to their ancient home with a stated reverence for the land,[9] economic pressures almost always took precedence over environmental ones, especially in a time before conservation of rivers was a significant concern anywhere in the world. The economic vitality of Israel came long before the country's environmental well-being.

In the first decades of the State of Israel, rivers, and environmental protection in general, were low priorities. With a life-and-death struggle afoot on every border, and a massive number of mostly impoverished immigrants from many countries to be absorbed, security and economic development were the focus of the government and society. The Jewish National Fund (JNF) was a steward of Israel's forests, planting millions of trees while creating shade and a means for anchoring the soil.[10] But there was no comparable governmental or not-for-profit organization that then saw itself as the advocate for Israel's waterways.

During those early years, there had been a consensus that the country's rivers should serve agriculture and the economy, which resulted in the pumping out of rivers upstream for irrigation before the water would be tainted by pollution. Meanwhile, downriver, most of Israel's coastal rivers—that is, those that run yearlong, flowing east to west, to the Mediterranean Sea—were allowed to be transformed into open sewer canals and municipal dumps. If manufacturers had to dispose of industrial or chemical by-products, they, too, would be directed to the nearest river for disposal.

In principle, Israel's rivers were protected by a series of laws, beginning with the Water Law of 1959,[11] and a comprehensive law on rivers and streams in 1965.[12] But, in practice, rivers were exploited for their practical value.

The Yarkon River is a case in point. Its death spiral began in 1955 with the diversion of its waters to feed the Yarkon-Negev irrigation plan—Simcha Blass's Phase II—that was later incorporated into a larger system when the National Water Carrier opened in 1964.[13] Urbanization also played a part in the river's decline. With little other use for the small amount of water still flowing along the seventeen-mile length of the Yarkon, the growing cities and towns along its path were glad to have the river as a cheap conveyance in ridding themselves of municipal waste in an era before sewage was seen as anything other than a nuisance.

Blass, while heading the government department on water utilization in the early 1950s, expressed concern about overpumping of the Yarkon. He also presciently wrote that the Yarkon would suffer lasting damage if the amount of sewage entering it continued to grow and the volume of water available to push the sewage and naturally forming sediments downriver continued to shrink. Blass wasn't driven by environmental worries directly but by a pragmatic concern that despoiling the Yarkon would have unintended consequences, even possibly ruining the aquifer beneath Israel's longest coastal river.[14]

By 1988, after more than a decade of on-again-off-again government discussions about how to address the befouling of the Yarkon, a dedicated Yarkon River Authority was established to develop and implement a plan to rehabilitate the river. After a few years of modest activity, Dr. David Pargament, an expert in the integration and management of water resources, was brought in to manage the organization in 1993.[15] He has been there ever since.

Passionate about rivers, in general, and the Yarkon, in particular, Pargament is as much philosopher as hands-on executive. On first meeting, Pargament looks like a double for a department store Santa Claus, with a barrel chest, a booming voice, wire-rimmed glasses, and a broad white beard. Only his white ponytail puts him slightly out of character. "There is no real connection anywhere between the city and the watershed," Pargament says, referring to the large land area that drains into a lake, river, or ocean. "Everywhere in the world, people have been cut off from nature and most especially, the natural flow of water. Roads, trains, buildings, and homes cross the hydrological plane. The watershed and all of its tributaries were all once connected, but today they are all cut off from each other. The Yarkon River Authority is charged with reconnecting the watershed to the people."[16]

The organization that Pargament manages is, in effect, the Yarkon's advocate. In that role, he pushes back against those who would encroach on the river and, where he can, prevents development, or at least guides that development to meet the Yarkon's needs while also rehabilitating the river and its banks, and restoring the river's natural habitats.[17] "The best-case scenario for the Yarkon River Authority would be to get all of the river water back and all development to go away," he says. "The worst-case scenario is the complete opposite, where developers put the river into a cement canal. But neither of those is going to happen." As the voice of the river, Pargament asks, "What, realistically, can I get? I can get something in between where we define the amount of water that will flow and its quality."[18]

To make up for the water that is still taken from the Yarkon, high-quality freshwater is pumped into the Yarkon to assure a regular flow.[19] That is water that could be used for agriculture or other economic purposes that is now allocated to the Yarkon (and other rivers) for the health of the rivers and for the value it brings to society.

"We need to provide for the river and the parks and the recreational areas around them," Pargament says. "Israel is a small country that is densely populated, and we are especially packed in the center of the country surrounding the Yarkon. But the more there is overcrowding of people in an area, the more there is a need for parks and rivers and recreation, because people need access to nature and open spaces. Whether there is a drought or not, we need to be able to irrigate grass and trees. And we need to let water flow in rivers."[20]

Not every river in Israel has enjoyed a revival as has the Yarkon. But the Yarkon is a success story, even with caveats, exceptions, and painful memories like the aftermath of the Maccabiah bridge collapse.

Species of fish thought gone from the Yarkon are back, and the fish feed on mosquitos, keeping the pest population low for nearby Tel Aviv and other cities along its path. Birds dive into the river to pluck out a native kind of sardine and other fish. Plant life that had disappeared is growing there again today. The riverbank has become a favorite locale for runners, rapid walkers, strolling couples, and even picnicking families. Israeli kayakers store their boats in sheds a short walk from the water's edge. Rowing teams from Europe come to the Yarkon to train in winter when their rivers freeze over.

"Even if we still have work to do, we've created a Yarkon River that works for the broadest interests," says Pargament. "The environment is respected. Agriculture gets the water that it needs. And the public has a river to enjoy. Some of the other rivers in Israel are still in need of repair and have to be cleaned and restored, but the Yarkon is a model that works."[21]

A Lake in the Desert

Despite being called a river, the Besor only flows for about thirty-five days a year.[22] Following winter storms, rain washes from Hebron in the West Bank and travels swiftly downhill, skirting the ancient city of Beersheba, moving across the Western Negev and through the Gaza Strip, where the journey ends with any uncaptured water draining into the Mediterranean Sea. During most of these randomly drenching rainfalls from November to March each year, the Besor River is a conduit for the resulting winter floods for no more than a few hours at a time. The riverbed remains damp, maybe even muddy for a short period after, but, as with all desert rivers, the briefly engorged river is soon no more than a memory.

As Israel began its construction of roads in Beersheba and the surrounding Negev area in the 1960s, it found great quantities of gravel under the Besor River that was excavated during the spring and summer months for use in the foundations of the area's roadways. More than one hundred acres of mining pits of varying depths were created along the fifty-mile length of the river. Aside from being an eyesore, pools of static water would remain in those mining pits following winter storms and they became breeding grounds for mosquitos, plaguing the people of Beersheba and the surrounding area.

With the Besor River also serving as a conduit for human, agricultural, and industrial waste from both the greater Hebron and Beersheba areas, untreated sewage would often pool in the gravel mining pits, unflushed by the winter storms passing over them. The pits became a foul no-man's-land around the river and developed into dumping grounds for construction debris, discarded appliances, and even cars with no salvage value.

Every so often, an extreme winter storm would cause the river to overflow its banks, carrying with it an array of unpleasant remains to dry in the desert sun, where the debris would remain. The floodplain became the effective southern boundary of Beersheba, Israel's fourth largest metropolitan area.

In 1996, following one such flood, the government finalized existing plans to create eleven river authorities, separate legal entities patterned on the model of the Yarkon River Authority. Each of the river authorities would be responsible for the monitoring and the rehabilitation of a major river (relatively speaking) and its tributaries, covering, in all, thirty-one Israeli rivers and streams.[23] The Besor River and its Beersheba River tributary were overseen by the Shikma-Besor River Authority, and, in 1997, Dr. Nechemya Shahaf, an expert in the management of rivers and an economist, became its head.

For Shahaf, a long-range plan was needed, but from the beginning, he saw potential that no one else saw. While development around a river is often a precursor to the river's pollution and ruin, Shahaf saw it as the river's salvation. While he stopped exploitation like dumping or gravel extraction, he didn't take an "all-exploitation-is-bad" approach.

"It quickly became clear that the river authority was about more than just a river and the environmental issues that obviously went with that," says Shahaf. "When the authorities were set up, it was assumed an integrative approach would be employed, looking at the river from many angles."[24]

For Shahaf, that meant more than the necessary first task of cleaning the river and filling in the mining pits. Shahaf reimagined the city of Beersheba growing south into a new luxury neighborhood where the flood zone then stood. "I knew it wouldn't be easy," he says. "This is the part of town that everyone avoided. No [real estate] developer would invest money in an area with such a terrible reputation. But if I was right, Beersheba would have more than a new neighborhood. It would have a new image of itself." Shahaf believed that the boundary of the floodplain could be redrawn, expanding the perimeter of the city and the outer limit of development on then uninhabitable land.

In 2003, Shahaf completed a five-year master plan that had an odd element, especially given that Beersheba sits on the northern edge of the Negev desert. His plan called for the development of an enormous park—more than fifty percent larger than New York's Central Park—running along a five-mile stretch of the Beersheba River. To build the park, the river would need to be restored and its banks made strong enough to protect against a once-in-a-hundred-years' flood event. Making his dream yet more fanciful, the centerpiece of the Beersheba River Park, he said, would be a 364-acre lake, but in a location where no water was then available.[25]

Shahaf found the first of several key partners in Russell Robinson, CEO of the Jewish National Fund's autonomous US organization. Although historically linked with the Israeli "home office," Robinson has long steered his large American-based charity to partner with the Israeli group when it made sense and to act independently when he saw opportunities the Israelis weren't ready to adopt. Whether from the US or the Israeli JNF, or from JNF branches in countries around the world, it is hard to find a water-themed project in Israel—or any environmental one—that doesn't get expertise and funding from one or all of these organizations.

Robinson saw in the park a potential catalyst for a long-held dream of his own, one even more ambitious than restoring the Beersheba River, as worthwhile as he believed that was. He wanted to affect "dramatic change" in the Negev region, but there were limits to what philanthropy could achieve. For his desired change to occur, the Negev would need a larger business and tax base, both of which, he believed, would best begin with a doubling of Beersheba's population, then at around two hundred thousand people.

The park, thought Robinson, could be a piece of a much larger whole that would also require economic development, outreach to the region's poor (including its indigenous Bedouin), and the development of a strategy to get the millions of annual tourists to Israel to think of the Negev as an essential destination.[26] Robinson and the JNF-US Board began boldly with an initial multi-million-dollar pledge from the organization. Over time, JNF-US's gifts have grown to tens of millions for the river park and for Negev development.[27]

Along the way, the Israeli JNF joined in, as did Israel's national government, each with large contributions and challenge grants. A dynamic mayor elected in Beersheba in 2008, Ruvik Danilovich, became the public face of the project, cheering small milestones like the opening of grassy sections of the park and large ones like the inauguration in 2013 of Israel's largest amphitheater, which is within the park.

As with the building of other great municipal parks in the US and Europe, the Beersheba River Park will be in development for a long time. And like those artificial urban evocations of nature brought into a growing city, the hope is that the Beersheba River Park will help to anchor municipal life. Already, other cities in the Negev have begun utilizing their rivers as a tool of urban renewal, including the Bedouin city of Rahat, which is located about ten miles north of Beersheba.[28]

The lake at the heart of the Beersheba park is expected to open around 2020, with a standard depth of five feet—despite about three feet per year of evaporation caused by the Negev desert sun. It will be refilled sustainably from the same source that will be used to water the grass, plants, and the six thousand trees planted in the park.

Surprisingly, none of the winter-storm water that flows through the Besor and Beersheba rivers will be captured for the lake or the park's grounds. That water will be permitted to flow to where millions of gallons are grabbed in each winter rain by a storm-capture reservoir in the western Negev, with that water reused on crops near the Negev reservoir.

Instead of the water saved from the winter storms, Shahaf's plan for the lake and park makes extensive use of the wastewater from Beersheba's homes following three rounds of treatment and purification. "It may seem that the water that will be used for the lake and for watering the park could be used for agriculture instead, but not everything has to be so practical," says Itai Freeman, a strategic planner, formerly associated with the Beersheba River Park project. "With enough water available, it raises the question of how you define quality of life. Among the questions you have to ask yourself is, how far does someone have to go to lie in a green open space? How far does a family have to go to be able to sit in a park under a tree? These are questions about the quality of life. Growing more crops is important, too, but life is about more than just that."[29]

The vision of transforming Beersheba has already started to become a reality. A recent visitor to the southern part of the city, where a dump site once was and which is where the park will rise, noticed a billboard at the base of a high-rise apartment complex now under construction, a short distance away. The Hebrew-language sign announced something unimaginable just a few years ago: "Luxury apartments for sale overlooking the Beersheba River Park."

"Not One River Has Been Completely Restored"

Peering over the shoulder of every Israeli government official is the state comptroller, an independent bureaucrat whose position seems to combine the best features of an investigative reporter, a forensic accountant, and an ombudsman. With wide-ranging authority to audit government agencies and to investigate abuse, waste, and underperformance, the state comptroller provides a powerful check on the government, making its operations more honest and efficient.

One of the state comptroller's recent comprehensive reports was devoted to the status of Israel's river rehabilitation efforts. The document was a mixture of praise and criticism, with suggestions offered for improvement. While acknowledging measurable progress in every one of Israel's rivers, the state comptroller called upon the government to spend much more to speed up the process.[30]

Rivers are resilient, when given a chance. With enough time and enough clean flowing water, nearly every polluted river will enjoy a revival. But with economic activity surrounding every waterway, no river in Israel can or will be returned to its natural state. It will take human intervention to improve what human activity and abuse earlier made worse; but it will always be a balance between economic interests and environmental restoration.

While more spending and effort are needed, Israel's rivers have been the fortunate beneficiary of two macro trends: the rise of environmentalism and the development of a comprehensive national infrastructure for wastewater and desalination.

As in much of the rest of the world, environmentalism had been a fringe concern in Israel until recent decades. Beginning in the 1990s, environmental laws were passed that required safe disposal of industrial and chemical by-products.[31] With ever stronger enforcement of these environmental laws, former polluters found ways to either change their manufacturing techniques or to have wastewater treatment on site before discharging wastewater carrying harmful elements.

The decision to reuse treated municipal sewage for agriculture has also had a significant effect. Although the rivers were not the intended, or primary, beneficiary of Israel's wastewater policies—agriculture was—the effect of the universal treatment of sewage and its reuse has been to get what was going down the drain of every Israeli's kitchen and bathroom out of the rivers. For so long as the waterways were choked with oxygen-depleting organic material, fish and plant life could not survive.

In more recent years, Israel's increasing reliance on desalinated water for domestic use served to take pressure off of the rivers in their upstream stretches. With more manufactured water available from desalination, it was possible to have the luxury of taking less water out of the rivers. With more water flowing, the natural cleansing that healthy rivers enjoy was enhanced.

Equal in significance to new laws and infrastructure, was a new attitude. As the benefits of environmental policy began to be broadly felt and as each city observed another city's successful rethinking of its interaction with its river, a fresh outlook developed. Rivers went from being eyesores of negligible worth to integral parts of each community's visual and emotional landscape. With local residents being drawn back to the rivers for leisure and recreation, real estate developers also took notice. Neighborhoods created or restored near the once-avoided rivers added to the momentum of this virtuous cycle of revival.

But even with all of these changes in recent years, the damage done over a period of decades would not be quickly resolved. Despite acknowledging significant successes in the twenty years since the creation of a rivers' initiative in the Israeli Environmental Protection Agency, the state comptroller's recent report also complained that not a single one of Israel's thirty-one rivers and streams had been entirely rehabilitated from its source to the sea—including the Yarkon, Besor, and Beersheba rivers.[32]

"We Are Now a Water-Rich Country"

Senator Henry Jackson, on a trip to Israel in 1970, reportedly believed he was being hoaxed when brought to the Jordan River. After being assured that it was no joke, Jackson was supposed to have said that the river's universal renown was "an act of public relations genius." At a different time, Henry Kissinger supposedly said that the river had "more reputation than water."[33] Whether apocryphal or true, the comments of the senator and the secretary of state reflect a duality about Israel's most internationally famous river.

There is a Jordan River that is a place of inspiration, imagination, religious devotion, slave spirituals, and folk songs. That river—"deep and wide" and "chilly and cold"—is the one from which, as the song has it, "Michael rows his boat ashore." It was across the Jordan River that the Children of Israel crossed into the Promised Land after forty years of wandering after the exodus from Egypt, and it was in that same river that John the Baptist immersed Jesus. But the Jordan River seen by Kissinger and Jackson, and likely many disappointed visitors, is, in its longest part, a mostly shallow stream that can be forded at many points with a single running jump.

The Jordan River is also two rivers in another way. The upper part of the river—the section that aggregates tributaries from a watershed that reaches north, east, and west, and that flows into the Sea of Galilee—has high-quality water that, if it is polluted at all, carries deposits from cows upstream in Lebanon. The lower part of the Jordan River exits the Sea of Galilee only in controlled amounts. The river meanders south to the Dead Sea, picking up agricultural runoff, pollution from fish farms, and domestic sewage from Israeli and Palestinian communities along the way, as it declines in volume and quality along its route. In all, in both its upper and lower parts, the Jordan River, Israel's longest, is only 156 miles.

Politically, the Jordan River and its tributaries have been a source of conflict between Israel and its neighbors in separate events in the 1950s and 1960s. The first dispute, as recounted in Chapter 2, was resolved in 1954 with the mediation of President Eisenhower's special ambassador, Eric Johnston. This led to a de facto agreement on the sharing of the Jordan River water by Israel, Syria, and the Kingdom of Jordan, which gets its name from the river.[34]

The second dispute began with the start of a water-diversion project by Syria of one of the Jordan River's tributaries, which was intended to deprive Israel of one of its important sources of water. The Syrian project was almost certainly less about water and more about saber rattling for the domestic political benefits it would provide the Syrian ruler. In any event, it would have been prohibitively expensive to complete the diversion project and unfeasible from an engineering perspective even if it were to be built in a peaceful location, which it was not. Syria began construction, but quickly lost the support of Egypt, the most powerful Arab state. The project came to an end with a single attack in 1964 by Israel—a pointed notice to Syria that Israel could upend the river diversion any time it wanted, even after enormous Syrian expenditure of money and political capital.

Although Syria never formally abandoned the project, it became moot. The 1967 Six-Day War settled the issue with Israel gaining control of the Golan Heights, which it has held ever since as a strategic buffer. With the Golan Heights, also came control over the Jordan River's tributaries, a bonus for Israel and its water security. While the Golan Heights are still considered disputed territory, Israel would be unlikely to relinquish the valuable high ground without both security guarantees and clarity on water rights to the Jordan River.

More recently, the Lower Jordan has been a vehicle for cooperation and even peace between Israel and the Kingdom of Jordan. The two nations bordering the river signed a peace treaty in 1994 and normalized relations, including comanagement of shared water resources. But even before they formally had peaceful relations, the parties tacitly worked together to manage the Jordan River, a confidence-building act that "paved the pathway to peace between them."[35]

Following the peace agreement, the Kingdom of Jordan received a significant boost from Israel in the management and supply of its water resources. Israel agreed to provide the kingdom with about fourteen billion gallons of water each year from its own supplies. Further, having no significant storage facility for its water, the kingdom also worked out an arrangement with Israel to store its water reserves from the Yarmouk River—a Lower Jordan River tributary under the control of the Kingdom that defines its northern border with Syria—in the Sea of Galilee. Jordan is permitted to withdraw its stored reserves from that body of water at will.[36]

As positive as it is that the two countries have made their shared part of the river a peaceful international boundary, the environmental quality of the Lower Jordan isn't likely to significantly change anytime soon without a new approach. The Upper Jordan—a major source of Israel's freshwater—will continue to flow in volume, and to be a popular place for those who want to kayak in Israel's most challenging waters. In contrast, the Lower Jordan enjoys only a relative trickle due to the small amount of water permitted to exit the Sea of Galilee.

One new idea for the revival of the Lower Jordan comes from Ram Aviram, a former Israeli ambassador for international water issues and now a professor of water policy at Tel Hai College in Israel's Upper Galilee. Aviram believes the southern part of the river needs more water for its well-being. "Because of the controlled flow, there is now less than ten percent of the water in the Lower Jordan than was there one hundred years ago," Aviram says. His concept would revitalize the Lower Jordan from the southern end of the Sea of Galilee to about the beginning of the West Bank boundary.[37]

He has suggested taking highly treated wastewater from Israeli cities like Tiberias and Beit Shean that is now used for agriculture and letting as much as five billion gallons a year flow into the Lower Jordan. "A revitalized Lower Jordan could become a hub for recreation, tourism, religious experiences, and bird-watching," says Aviram. He also believes that this would be a boost for the Jordanian economy, which is good for the Kingdom of Jordan and also good for Israel, which is eager to see its neighbor prosper.

"With our access to desalinated water, our efficiency in agriculture, and our reclaimed water," Aviram says, "we can afford to divert water for ecological purposes. The more water that flows in the Jordan River, the healthier it will be. We are now a water-rich country and we can act like it with the Jordan as we do elsewhere."[38]

The water permitted to flow in Israel's rivers and the large man-made lake adjacent to a desert in Beersheba is a metaphor for the transition Israel has enjoyed. Once, like water-stressed countries everywhere, Israel overdrew on its natural water resources. Now that Israel enjoys the abundance that it does, it can afford to renew its rivers, develop water-based recreation, and be creative with its water assets.

For those nations that do not soon develop answers for water scarcity, a likely further consequence will be environmental degradation. Aquifers will be overdrawn, rivers will grow ever more polluted, and fish and wildlife will die, among other unhappy outcomes. By having a surplus of water, rivers can flow and the standard of living can rise along with the quality of life.

"An Integrated, Flexible Water System"

Despite its name, the Sea of Galilee isn't a sea. It is a lake. Long Israel's largest single resource for freshwater and a major contributor to the National Water Carrier, it is also a center for recreation, a destination for tourists and Christian pilgrims, and home to a small fishing industry. Today, far more water is derived from desalination than from the lake, and even before desalinated water was available, more water was taken from Israel's several aquifers each year than from the Sea of Galilee.

Even so, the Sea of Galilee has played an outsized role in the nation's consciousness as a barometer of the country's water concerns, and at times even as a reflection of the national mood. "When Israel suffered droughts," says Shimon Tal, a former head of Israel's Water Commission, "everyone knew the level of the Sea of Galilee. It would be announced on the nightly news and in the daily newspapers. If it seemed to stabilize, people would be happy. If it approached the lower limit—a number everyone seemed to know—people would grow nervous about the future, and be extra careful with the water they used."[39]

The Sea of Galilee is geologically part of the Syrian-African rift, a deep depression running thousands of miles, with its lowest point being to the Sea of Galilee's south at the end of the Lower Jordan River at the Dead Sea, also a lake. The Sea of Galilee sits nearly seven hundred feet below sea level.

The rise and fall of the Sea of Galilee's water level—a swing of about fifteen feet—marks the range of health for the Sea of Galilee. Above an imaginary upper red line, flooding of the area around the lake is likely, and below it, there is a fear that permanent ecological damage could be done to the lake. An excess of water can be adjusted quickly by increasing the flow into the Lower Jordan River, which is south of the Sea of Galilee. But in drought conditions, the whole country was primed to keep its eyes on the lower imaginary red line. No water officials wanted to test what it would mean for the country's water future if the lake were pumped far below the consensus figure of possible hazard.

Aside from the risk to the lake, there was also a risk of overpumping Israel's aquifers. The two main aquifer systems run along north-south routes. One, adjacent to the Mediterranean shore, called the Coastal Aquifer, is a shallow resource. The other, running under the Samarian Mountains and Judean Hills, and to their west, is called the Mountain Aquifer. If the Coastal Aquifer is overpumped, seawater will enter. The Mountain Aquifer is also at risk of overpumping and of pollution entering the aquifer, but not of being ruined by seawater.

Unlike with an underground aquifer, which is a closed system, evaporation is a problem for surface water like a lake. This is even more so in a hot climate below sea level with few cloudy days during three seasons of the year. Evaporation claims nearly as much water each year as is diverted from the Sea of Galilee for national consumption. In a normal year, five feet of water from the surface of the lake is lost to evaporation.

After a few years of drought, the shore of the Sea of Galilee recedes so far that archeological treasures—including a fishing boat believed to be from the time of Jesus—have been found on the exposed shore.[40]

Today, even with a growing national population and a vibrant economy, the Sea of Galilee is stabilized, with withdrawals limited to an amount that will keep the lake between the imaginary red lines that indicate its health. Like Israel itself, Israel's freshwater lake is now largely immune to weather fluctuations, and even to droughts of a year or two. It is likely that the fears of the lower red line of the Sea of Galilee will soon fade and become a generational memory of a bygone era.

Still, because today the Sea of Galilee is responsible for a smaller but still important portion of the country's drinking water, comprehensive scientific monitoring of the lake's water continues. Microbiologists and chemists constantly check the water for foreign bodies, water clarity, salinity, and an ever growing list of categories. Believing that data collection will yield valuable trend information, Mekorot has been tracking the Sea of Galilee's water for decades.[41] With all this information in hand, Mekorot can identify microbes in the water, pesticides, incipient algal blooms, and other threats to the water supply or anomalies from a seasonally adjusted benchmark before the drinking water supply is threatened.[42]

Monitoring the lake has also resulted in a build-out of Israel's national water infrastructure. After certain micro-pollutants were found in the 1990s while routinely checking for water quality, a decision was made to build the Eshkol Filtration Plant, in Bet Netofa, about twenty miles west of the Sea of Galilee. The facility—among the largest filtration plants in the world—is a high-tech center monitoring and filtering the water coming from the lake. The control room of the Mekorot facility is modestly staffed, but filled with colorful monitors to alert the engineers on duty of any change in water quality.

While Israelis don't speak openly about defenses against terror attacks with outsiders, it is clear that the staff at Eshkol Filtration would know in moments if toxic materials entered the water supply, whether from terrorism, an accident, or natural causes. Likewise, if algae or some unwanted foreign body entered the water, the monitors would promptly send an alert.[43] As a result of the monitoring and filtration, Israel's citizens enjoy tap water of a quality normally associated with more expensive water choices. "I know that many people think bottled water is safer than tap water," says Mekorot aquatic biologist Dr. Bonnie Azoulay, "but under the microscope, there is no difference between Israeli bottled water and water from the tap. The tap water in Israel is clean and safe, and it is what I drink."[44]

Beyond monitoring, Israel also makes a large effort to maximize the quantity of water that falls into the Sea of Galilee and its surrounding watershed. Since the late 1950s, Israel has been seeding clouds with silver iodide in the winter months to enhance the amount of rainfall.[45]

By the 1960s, Israel had put a lot of resources into testing rain-cloud seeding and developed world-renowned expertise in how and when to seed. It is believed that cloud seeding may add as much as eighteen percent to the rainfall over the Sea of Galilee watershed and about ten percent to what falls on the lake itself. The technique may be adding as much as ten billion gallons of water a year to the lake. At a cost of only about $1.5 million for the annual Mekorot cloud-seeding operation, this is very inexpensive water.[46]

Another benefit of the diminished reliance on water from the Sea of Galilee is a healthier profile for Israel's water as a whole. The Sea of Galilee sits on a layer of salt, and this salt pushes its way into the lake's water.[47] There were also saline springs that leaked salt-rich water into the lake until a challenging water-diversion project rerouted the salty water to the Lower Jordan.[48] Because of these salty intrusions, Sea of Galilee water has always had a high-salt content.

With approximately one-third less water being pumped from the lake each year since wastewater treatment and desalination have been added to the national water mix, there is that much less salt being consumed in Israel—and that much less finding its way into agricultural use after wastewater treatment. The extra water left in the lake helps to stabilize the ecology of the Sea of Galilee, with fewer wide swings in the level of the lake induced by fickle weather patterns.[49]

"Operationally, functionally," says Shimon Tal, the former water commissioner, "we have turned the Sea of Galilee into a reservoir. We can call on it when we want and we can let the water accumulate for a dry period, too. We have more water for nature and we can increase the flow of the Jordan River. We can make less use of expensive desalinated water, or we can give our aquifers a year or two to recharge by making less of a demand on them. The Sea of Galilee is still an essential part of our water supply, but now it is a part of an integrated, flexible water system that is mature and resilient. No one lies awake at night anymore worrying about the lower red line of the Sea of Galilee."[50]

Drip irrigation saves enormous amounts of water over traditional irrigation techniques such as flooding the field or using sprinklers. By only dripping water at the roots of the plant, little water is lost to evaporation. As important, drip-irrigated plants grow more robustly, routinely increasing yield by 100 percent and often far more. Plants can also be given nutrients and fertilizer by drippers, preventing nitrogen from being washed into waterways and reducing the amount of chemicals on the fields. (Netafim)

Lush, drip-irrigated fields found in Israel's Arava Desert demonstrate how worthless desert sands can be used for growing a wide array of crops. As the world becomes ever more insecure about adequate food supplies, it will likely have to turn to currently undesirable land to grow its fruit and vegetables. The sand serves as an anchor for the roots, and the plant can grow without being bothered by traditional pests found in more humid growing areas. (Netafim)

India is now the world's fastest-growing market for drip-irrigation equipment. As of 2014, Indian farmers, like those pictured here, had more than five million acres that utilized surface and subsurface drip-irrigation lines. With a need to conserve water resources and produce more food per acre for the country's fast growing population, the Indian government lends money to farmers to encourage their switch from traditional—and wasteful—flood irrigation to drip irrigation. (Netafim)

Naty Barak was a farmer at the kibbutz that developed drip irrigation. He reluctantly became an executive at Netafim, the first drip irrigation company. Today, he is an evangelist for the economic and social value the irrigation process provides in saving water, increasing food supply, lowering use of carbon fuels, reducing gender inequality, and aiding poor farmers in less-developed countries. He is shown with the Queen of Sweden after accepting the Stockholm Water Prize for Netafim in 2013. (Netafim)

After deciding in the 1970s to make widespread use of treated sewage for agriculture, a new national infrastructure parallel to the freshwater system was developed and built, from treatment plants to separate pipes to reservoirs to hold surplus water. The network of reservoirs pictured here, like the hundreds of treated water reservoirs found around Israel, was built with financial support from the Jewish National Fund. This Negev Desert reservoir can hold nearly two billion gallons of reused water for desert farming. (JNF)

To safeguard Israel's water from a range of threats, Israel has among the world's largest and most high-tech freshwater filtration systems. Opened in 2007, the Eshkol Central Water Filtration Plant, near Nazareth, purifies all of the water coming from aquifers or the Sea of Galilee, and also monitors that water for quality, foreign bodies, and toxins in a sophisticated system able to instantly stop the flow of tainted water and replace it with water from other sources confirmed as pure. (Mekorot)

While most major cities have wastewater treatment plants, Israel's Shafdan facility is unique. Aggregating all of the sewage in the greater Tel Aviv area, the Shafdan sewage is treated and then percolated through layers of purifying local sand. The water is then pumped out of a dedicated aquifer and transported to Negev Desert farms via a dedicated pipeline. With 85 percent of Israel's sewage treated and reused for agriculture, Israel has a major new source of water that complements natural water sources. (TAHAL)

Shy and modest Sidney Loeb is the unsung hero of the reverse osmosis desalination revolution. Inventing a new way to remove salt and minerals from salty water as a researcher in California, Loeb moved to Israel where he became an engineering professor and a local celebrity while conducting tests that advanced his invention. Shown here with his Israeli wife, Mickey, in 1970, Loeb earned all of $14,000 from what grew into a global multibillion-dollar industry. (Mickey Loeb)

Former Stalin political prisoner Alexander Zarchin convinced Israeli Prime Minister David Ben-Gurion in the 1950s to set up a government agency to test his concept of desalinating seawater by freezing it. Although it ultimately failed to produce drinkable water at an affordable price, the Israeli government used the accumulated know-how from this first major desalination effort to launch research that later led to Israel becoming a powerhouse in desalination. (IDE)

Israeli and U.S. leaders Levi Eshkol and Lyndon Johnson developed a warm relationship over water. Both men had backgrounds from water-deprived farming communities and both dreamed of abundance of water as a vehicle for economic development and peace. Johnson had a special interest in desalination and sought out a partnership with Israel to further research and development. The normally talkative Johnson listens attentively to Eshkol during the latter's 1968 visit to the LBJ ranch, Johnson's home. (Yoichi Okamoto/LBJ Library)

Soreq, the world's largest seawater desalination plant, is located about a mile from the sea. The Israeli facility manufactures 165 million gallons of fresh water a day, or 7 million gallons an hour. By making use of a proprietary algorithm that accesses electricity at dips in price, Soreq produces the least expensive desalinated water anywhere—a fraction of a penny per gallon. Israel's Mediterranean desalination plants now provide the equivalent of 80 percent of Israel's household water. (IDE)

One beneficiary of Israel's new abundance in water has been the country's environment, including its rivers. With alternative sources of water now available, less river water is needed and more is permitted to flow, keeping the rivers healthier. Dr. David Pargament, pictured here, heads the Yarkon River Authority, where he serves as a guardian of the river, also pictured here. If Israel's rivers were once badly polluted, today they have been revived and now serve as a place for recreation. (Yonatan Raz/David Pargament)

Whether in desalination, membranes, drip irrigation, water security, or valve-to-control-room communications, among many other fields, Israel has used technology to improve water efficiency at ever lower costs. In the process, Israel developed a new export industry that now sells the technology around the world. In 2013 President Shimon Peres presented an award to Amir Peleg, a technology entrepreneur and founder of the leak minimization company TaKaDu. (TaKaDu)

Helping to improve Africa's water supply has been of great interest to Israel since the 1950s, with Israeli water development officials present in nearly every sub-Saharan African country now or in the past. Innovation: Africa, an Israeli NGO, continues that tradition. Founder Sivan Ya'ari helped to develop a remote-controlled, solar-powered water pump that provides clean drinking water to African villages that is run from the NGO's offices in Tel Aviv. (Innovation: Africa)

Beginning in 1960, Israeli water engineers helped run Iran's water system. All of the many Israeli personnel left the country in the days before the 1979 Islamic revolution. Since that time, Iran's water supply has grown to be among the world's most imperiled. Israeli water engineer Shmuel Aberbach, second from left, at Hasht Behesht Palace, Isfahan, Iran, in 1963 traveled all over Iran training Iranian engineers, some of whom were exiled or executed after the revolution. (Shmuel Aberbach)

Professor Arie Issar, right, headed the Israel water development team in Iran. His work was well known by the country's leadership, and appreciation was often expressed by both senior Iranian government officials and local water administrators in all parts of the country where Issar and the Israeli hydrologists worked. The Shah of Iran, left, visited Issar and one of the water-drilling projects in the late 1960s to express his gratitude for the work of the Israelis. (Arie Issar)

Several Israeli government-owned companies worked widely and openly in Iran's water sector for nearly twenty years, virtually never encountering anti-Israel or anti-Jewish sentiments. The Israeli water engineering company TAHAL (an acronym for the Hebrew words "Water Planning for Israel") led efforts to rebuild the water systems in Qazvin (Ghasvin) following a devastating earthquake there in 1962. The TAHAL office sign from the 1960s is shown here in Farsi and English. (TAHAL)

The Red Sea–Dead Sea project will bring desalinated water from the Red Sea to Israel's Arava Desert and, in exchange, Israel will transfer water of its own to the Kingdom of Jordan and the Palestinian Authority. The usually discarded brine from the desalination process will be transferred to the Dead Sea to slow its evaporation. Perhaps best of all, Israel, Jordan, and the Palestinians all become more interdependent. Water officials Silvan Shalom of Israel, left, and Hazem Nasser of Jordan are shown, in 2015, holding signing pages for this innovative agreement. (Israel Government Press Office)

California and Israel share similar climates as well as rapidly growing populations and economies. Since a long drought has hit California, water resources have become challenged. Among the steps taken have been to reach out to Israel to identify ways that California could learn from Israel's water practices. In March 2014, California Governor Jerry Brown, left, and Israeli Prime Minister Benjamin Netanyahu signed a cooperation agreement that encouraged government, business, and academic partnerships between Israel and California for the smarter use of water. (Leah Mills/Polaris Images)

Israel has long shared its water technology to benefit other countries but also to deepen Israel's political and commercial connections with them. Among the reasons China decided to establish diplomatic relations with Israel in 1992 was that Israel could provide help in improving China's growing water problems. Shown here, Chinese Premier Li Keqiang, left, welcomes Israeli Prime Minister Benjamin Netanyahu to China in May 2013 for discussions that led to Israel renovating a Chinese city's water system. (Israel Government Press Office)

PART III

THE WORLD BEYOND ISRAEL'S BORDERS

Eight

TURNING WATER INTO A GLOBAL BUSINESS

There isn't a scarcity of water. There is a scarcity of innovation.
—Amir Peleg, Israeli entrepreneur

ODED DISTEL LOOKS like he works in the entertainment industry. With his shaggy hair, wire-rimmed glasses, and warm personality, he's the opposite of the stereotyped image of a government bureaucrat. He laughs easily and has smile lines at the corners of his eyes. Born to two Holocaust survivors and raised in Israel, he studied business and went to work in a government ministry focused on trade issues, where he has been his whole career.

With the Athens Olympics scheduled for 2004, Distel was posted to Greece in 1998 to do the usual work of a trade representative, and especially to help Israeli companies sell their renowned homeland security know-how to the summer games. A year or two before the end of his term in Athens, he began reading about the first glimmers of the Cleantech revolution, a now obvious idea that energy, water, and sewage can be managed in a more efficient and environmentally friendly way. Distel immediately thought Israel could play a leading role, especially in water. He focused on Israel's significant skills and technology in areas as disparate as drip irrigation, wastewater treatment, and desalination, but suspected there was a lot more beyond that.

Distel did what all good government officials do: He wrote a memo. In just a few pages, he laid out the case as to what Cleantech could mean for Israeli business, in general, and how the Israeli government could help launch a big Israeli presence in the water industry, in particular. By their nature, bureaucracies are antagonistic to new ideas, and the first response to his memo was a turf-focused one. If he was talking about clean water, he was told, this was the jurisdiction of the Ministry of Environment and not the Ministry of Trade. No one he reported to was interested.

After Distel's return to Jerusalem in 2003 and in a new position in his ministry, he started lobbying for his idea. He contacted senior people in the bureaucracy, and tried to persuade them that this was an idea whose time had come. Israel had all kinds of special abilities in water, he said, and the world would beat a path to Israel's door—if they only knew that Israel had developed this mastery.[1]

There must have been something in the water.

At the very time that Distel was reaching out to his bosses about his water idea, Baruch "Booky" Oren had similar thoughts. Known universally in Israel by the nickname his mother had given him when he was four days old, Oren had recently become the head of Mekorot, the government-owned national water utility. Throughout his career, Oren had been a disruptive thinker, or what the corporate world often calls a change agent. He held a variety of jobs and in each one he would try to reimagine not just his position, but his organization, too.

Years earlier, after Oren's compulsory army service, he first studied biology and then business administration. One of his professors, just back from a long stint teaching at Wharton, was eager to start a consulting company in Israel and made two unorthodox choices in picking him and another unusually promising student as his business partners.[2] Oren was seemingly on top of the world.

Soon after, his baby son became ill and Oren and his family picked up and moved to New York for advanced cancer care. For years, they all lived in limbo. When his son died just before his thirteenth birthday, Oren took a position in Israel as a marketing manager at an Israeli software company. That somehow led to a job as head of business development at Netafim, the company that developed drip irrigation.

In 2003, the pro-business, pro-innovation government of Ariel Sharon came to power. Oren was recruited to become chairman of Mekorot.[3] As was his philosophy while working in his previous positions, Oren wanted to take a change-averse, once innovative organization and try to transform it. What, he asked himself, was the reason that Israel had a successful high-tech industry? He concluded it was primarily attributable to the Israeli army, a force that had to do "more with less" against an array of threats by learning how to make technology work in its favor. If Israeli high-tech could be born with the military as the driver, why couldn't a technology-driven water utility rethink the world of water?

Oren sent out the word to his Mekorot staff that he wanted to be informed of every problem its engineers faced. A mostly skeptical company produced a list that Oren shared with the world of inventors and entrepreneurs. He made a remarkable offer: If they could help to solve Mekorot's problems, they would retain the intellectual property rights to the solutions and benefit from the commercial exploitation of it. They could form companies around those ideas, and Oren and Mekorot would help incubate those companies' development.

To encourage the entrepreneurs, Oren proposed Mekorot as a beta site for these just formed companies' products and services. He offered them seed capital from Mekorot and even to be the product's first customer. To make sure this initiative would get off the ground, he also promised to help these new companies market their inventions to other utilities around the world. Like the Israeli army, which was mostly looking out for its own needs, Oren wasn't focused on building the inventors' businesses, even if he was happy for them to succeed. His hope was that their ideas would jump-start the use of technology at his company.

"If you are a utility, you can't do anything about your unionized labor costs and your fixed overhead is what it is," says Oren. "My problem was the same as every other utility in every country. The only way any of us can save is through increased productivity driven by technology."[4]

The challenge was that most utilities were as conservative as was Mekorot before Oren arrived. With water prices mostly controlled by state or local governments,[5] utilities don't have a lot of incentive to take risks with innovation.

But whether it is local or global water problems that are to be addressed, innovation will need to be a central part of the solution. Cash-strapped governments have no appetite for major new water projects at the very time that demand for water is growing all over the world. With utilities, agriculture, and engineering departments all inclined to continue doing what has been done, a new culture that encourages innovators and speeds adoption of new ideas is a key element of fixing water problems before they become crises.

"There's a lot of 'If it ain't broke, don't fix it' thinking around water," Oren says. "Technology is the poor man's way of getting a lot more out of your water system. But first, utilities have to go from thinking of themselves as just being a source of water to being part of a high-tech solution."

As a "big idea" guy, Oren soon decided that this was far bigger than just Mekorot. Why, he wondered, couldn't Israel turn its water expertise into an export industry in the same way that the army-inspired Israeli high-tech industry was a key driver of the Israeli economy? Utilities needed help, but so did agriculture and consumers and food companies and industry. Booky Oren thought he had stumbled on Israel's next great business idea.[6]

Bigger Than Biotech and Telecom

Although rarely thought of as such, Israel is a former Third World country, what is today called a less-developed country. Like its neighbors and most of Africa and Asia, Israel achieved statehood in the period of decolonialization bookending World War II. Egypt became independent in 1922, Jordan and Syria in 1946, India and Pakistan in 1947, and Israel in 1948. That Israel went on to become a leader in technology, life sciences, and defense, among other industries, was not preordained. Some of it was tied to luck, but some also to smart choices.

Every major part of Israel's economy played a supporting role in the development of Israel's water industry. Today, most water start-ups come from private sector laboratories, but in Israel's early years they were more likely to grow from a farm cooperative or even have been established as a government entity. But the development and growth of Israel's water commerce would not have occurred without sophisticated agriculture, manufacturing, financial services, and technology sectors. These all grew independently, but ultimately came together to help create the business of water in Israel.

At first, agriculture played a larger role in the pioneering spirit of the *Yishuv*, the Jewish community in Palestine, than it did in its economy. After millennia of exile in which Jews were mostly landless by law and found professional expression as rabbis, artisans, merchants, or peddlers, the return to Zion was accompanied by a call for a New Jew—men and women who would redeem both themselves and their ancient homeland by working the soil. The pioneer slogan "We came to build and to be rebuilt by it" captures this ethos. Nearly all of Israel's early political and military leaders spent at least a part of their formative years on a farm. Yet, despite this apparent centrality, at no time did agriculture make up more than thirteen percent of the economy.[7] Today, it is responsible for less than three percent of GDP.[8] Nevertheless, agriculture plays an important part in the country's psyche and as a source of national pride. Israel's agricultural economy, moreover, has become a highly technological arena focused on efficient use of water.

The manufacturing sector in the Land of Israel got a large boost when the British army in World War II discovered it was cheaper and safer to get products it needed for nearby troops from Jewish-owned factories than to ship them from more traditional manufacturing hubs located far away.[9] Factories were born or greatly expanded in many categories including garments for uniforms, food and beverages, and even in medicines purchased by the British army from a then small company called Teva, which is today Israel's largest publicly traded company. When the war was over and the new nation was created, these and other industries in the young Israel had a sophistication that had been accelerated by round-the-clock shifts and rapid expansion stimulated by the large British demand. By the end of the war, manufacturing was a third of the *Yishuv*'s economy.[10] So, when the time came to begin manufacturing sophisticated water devices, the local manufacturers knew how to do it.

The largest economic segment, even before independence, was the service sector. More than half of the economy was tied to education, medicine, research, and financial services.[11] Although no one knew it at the time, these (and related) business categories would put Israel in a good position for a world whose affluent nations were moving to service economies.

The young state was very poor, rapidly absorbing mostly penniless immigrants. From May 1948 until the end of 1952, the population more than doubled.[12] Food was rationed and the economy was reliant in a large way on foreign aid, German reparations, and the donations of world Jewry.[13] These funds were largely allocated wisely. Among other uses, new communities were inaugurated, national infrastructure was built, institutions of higher learning were born or expanded, research facilities were created, and a modern army was empowered. Reports of financial corruption were rare and incidents of officials diverting other than petty amounts of government or donor funds for their own use were essentially unheard of.[14]

All of this investment in research, higher education, and infrastructure proved to be wise when, beginning in 1989, about one million Soviet Jews came to live in Israel as the Soviet Union withered and died.[15] These immigrants, like the wave of German Jews who arrived in the

Land of Israel in the 1930s, were largely well educated and technologically adept.[16] The combination of existing institutions and a large body of technically sophisticated native and immigrant Israelis—at the very moment that the tech revolution was changing the world—offered Israel the opportunity to be a leader in this revolution.

Israel today is often called the "Start-Up Nation," after the bestselling book of the same name by Dan Senor and Saul Singer. The book identifies several reasons for Israel's success in tech industries, including an entrepreneurial culture, a military that identifies and trains the best technology minds, and the global perspective that Israel's regional isolation has spurred.[17] Another key reason is that Israel is routinely among the top spenders, per capita, on R&D. For example, in 2013, on R&D, Great Britain spent 1.6 percent of GDP; the United States, 2.8 percent; Germany, 2.9 percent; and South Korea, 4.15 percent. Despite also having the highest GDP defense burden of any developed country,[18] Israel spent 4.2 percent of GDP on research and development.[19]

All of this spending has yielded profound results. Today, more than 250 global companies have facilities in Israel. For many, like Google, Facebook, Apple, Intel, Microsoft, IBM, Hewlett-Packard, Motorola, General Electric, and Dell, their R&D center in Israel is their first, largest, or only R&D center outside of their home country.[20]

These R&D skills and this entrepreneurial mind-set have been applied to water, too. Until the last few years, the old paradigm in water had been that if you needed more of it, just add more capacity. Drill more, pump more, and add more pipes. The new paradigm is to increase the efficiency of water use—to make every drop count and to reuse each drop as many times as possible.[21] To change thinking about water from a resource-shortage problem to a scientific-innovation one, especially in conservative industries like agriculture, utilities, and infrastructure, you need entrepreneurs and a culture that challenges conventional wisdom.

The global water industry has annual sales of $600 billion, making it bigger than biotech and telecommunications and only a touch smaller than the global pharmaceutical industry.[22] Seventy-five percent of those sales are generated by what might be called "old" or "dumb" water: valves, pipes, pumps, and most of what utilities do. The other twenty-five percent in revenues are for high-tech products—areas like technology, desalination, membranes, leak minimization, drip irrigation, filtration, water security, and valve-to-control-room communications—which are the future of the water industry.[23] In each of those areas, Israel excels.

A Government-Favored Industry

Booky Oren knew that he'd need help in selling his new idea. Turning Mekorot into a technology friendly company and encouraging water exports was just the start. He dreamed of more.

Ever the change agent, Oren wanted the government of Israel to do something it philosophically opposed: to create a favored industrial sector. He was sure that Israeli water could grow into a large and profitable export industry, but it would only have the global impact he imagined if the Israeli government would give it a helping hand.[24]

He reached out to Ori Yogev, a former director of the all-powerful Budget Department in the Ministry of Finance. Coming from a water family—his father, David Yogev, headed the water-planning department at the Ministry of Agriculture for decades and had been a moving force in getting purified wastewater reused on a national scale—Ori understood and shared Oren's vision.[25] But there was a problem, perhaps insurmountable.

Yogev's former colleagues at the Ministry of Finance were philosophically opposed to government intervention in business and they were the ones who would have to make the decision to create a government-favored industry. If a business sector had huge potential, the Finance Ministry officials would say it would thrive on its own. Further, they argued, if the government were to lend a helping hand to an industry, all industries should be looked at. Maybe medical devices or aerospace were more worthy. Or maybe the Israeli government should try to change the national corporate culture altogether from start-ups to a preference for a few very large companies. There was no reason that water should be given any preferential treatment—assuming the government even wanted to get into the business of picking winners and losers.[26]

The best way to get the government to come around, Yogev concluded, would be to change attitudes about the potential of water exports. The water business would have to be seen as like none other. Oren and Yogev recruited Ilan Cohen, a former director of the prime minister's office; David Waxman, a former head of IDE, the desalination company; and a prestigious group of others.[27] Together, in 2005, they formed an advocacy organization called Waterfronts, which Yogev headed on a volunteer basis.[28] The group managed to get the government to consider the idea and convinced them to hire an outside consultant to analyze the global competition, the likelihood of success, and the potential size of the new, supported industry.[29]

The process succeeded. Research showed that a third of the world would be living in water-stressed regions by 2025 if action weren't taken soon to head that off. Rich and poor countries alike would be affected. The export potential was significant; certainly many billions of new revenue.

Israel's Finance Ministry was also impressed to learn that the global water industry was a huge but fragmented market. Unlike other categories that might also lay claim to favored government status, in water, entry wouldn't be blocked by a handful of corporate giants who could stomp out competition. But without the government boost that Oren had been pushing for, big overseas markets might never even notice the Israeli companies. Most concerning, by failing to move soon, another country might jump in, and the opportunity could be lost altogether.[30]

In a lucky break, just as this issue was being considered by the Finance Ministry, the government appointed a new professional director of that ministry: the departing head of the Trade Ministry, which, under his direction, had come to support government intervention in water. With the stars now aligned, the government blessed the creation of a new organization with government funding to focus on supporting Israel's water industry.[31]

Oded Distel, the Israeli trade official who wrote the first memo supporting water exports during his posting in Greece, had continued to press his case. Fittingly, Distel was asked to run the new government agency.[32]

Socialist Farms and Capitalist Results

A handful of Jewish socialists and communists were the first ones in the Israeli water industry. Ironically, it was fervent anticapitalists who ended up spawning several of Israel's most significant export businesses.

In the main, the men and women who came to the Land of Israel in the early part of the last century had not been farmers in their native lands. Even if they had been, the hardship and deprivation they found in their new homes was so extreme that the only way they could work the land was by banding together. By farming as a collective, they could share the burdens of work and security, mitigate the loneliness and frustration, and have partners to pick up for them when malaria or other diseases struck.[33] Zionism and socialism brought them to their new home. And so, the kibbutz, a Jewish communal farm, a revolutionary idea, was born. Everyone worked together and all ownership—even down to personal clothing—was communal.

As the decades passed, and even after the hardships diminished and the State was created, more collective farms were born. In all, nearly three hundred were established, routinely placed along the country's insecure borders.[34] These were farms, but also outposts to serve as an early warning system and a deterrent against terrorist incursions.

While these collective farms continued to be engaged in agriculture, beginning in the 1960s many of them built factories. At first, these were only staffed with their own members. Later, rather than hiring workers, they would give away parts of successful kibbutz companies to other collective farms, but ultimately, they all came to hire outside workers and even outside managers, transforming kibbutz life and softening the ideological edge of the founding generation.

Logically, many of the kibbutz factories developed products tied to what these farms knew best: agricultural products, especially ones connected to water and irrigation. Over time, several of these companies not only grew to be larger than their agricultural forebears but became global in scale.

One example can be found at Kibbutz Amiad, a collective farm in the Upper Galilee. During the early years of the state, all of the farms in Israel were bedeviled by clogging in irrigation hoses. The small holes in those hoses would often get jammed with dirt preventing the water from reaching the crops. As a result, plants would die and part or all of a harvest would be ruined.[35]

In the late 1950s, a farmer on Kibbutz Amiad with sophistication in engineering developed a system that used hydraulic power to unclog the hoses. It used neither electricity nor chemicals. It was a short jump from that field-tested invention to forming a kibbutz-owned and -run factory making hydraulic filters. In 1962, Amiad Filtration Systems Ltd. (now called Amiad Water Systems Ltd.) was formed.

Today, Amiad is listed on the London Stock Exchange. The company has more than four hundred employees, and sells its products in eighty countries. It recently acquired Arkal, a similar business started at Kibbutz Beit Zera, a kibbutz down the road from them, that was, oddly enough, its major global competitor. Amiad spends millions of dollars each year on R&D,[36] and still develops products for agriculture, but also for offshore drilling operations, desalination plants, and for filtering the ballast water of commercial ships, among other areas.[37]

More dramatic, Kibbutz Evron started a water business intended to save the lives of farmers. With kibbutzim and other farms mostly located along unfriendly borders and with hostile states on the other side of those borders, a farmer could be shot by snipers when going to the field to shut a water valve used for irrigation. But a valve that could be opened and closed remotely would allow ongoing irrigation of the fields, even during periods of tension or conflict.[38]

In 1965, nearly thirty years after Kibbutz Evron's founding, the kibbutz decided to expand beyond agriculture. A member there learned of an inventor who had developed a valve connected to a meter, and the kibbutz bought his invention, a product designed to shut the water flow using a hydraulic, nonelectric, process when a preset amount of water had gone through the meter. After spending a few years assuring that the product could be made to work reliably, the then single-product company marketed it under the name Bermad, combining the Hebrew words for faucet or tap (*berez*) and measure (*mad*).[39] With the Bermad device, the precise amount of water needed could be used and not an unintended drop wasted, a significant boon for farmers in water-constrained parts of the world.[40]

Global marketing of the Bermad device began in 1970. Demand for the product was great and led to Kibbutz Evron expanding its irrigation-product offering and also providing control valves for utilities and fire protection. The products are sold in eighty countries. Bermad has six hundred employees. The company is owned exclusively by Kibbutz Evron and another collective farm, Kibbutz Saar, taken in as a partner.[41]

Today, kibbutz factories can be found in several business sectors, but a large number of these manufacturers work in water realms far removed from the interests of farmers. Kibbutz Dalia is the home of Arad, makers of automated water meters. ARI was begun by Kibbutz Kfar Haruv and sells high-end metal valves for water utilities and other uses. Kibbutz Kfar Blum's Galcon provides inexpensive lawn-watering technology for homes and public parks. Kibbutz Maagan Michael began Plasson, a plastics molding business, looking to replace the metal drinkers found in chicken coops that were expensive and which would oxidize due to the bird waste.[42] But the company expanded far beyond its origins. It now employs more than twelve hundred workers and has products that include dual-flush toilets and a catalog of plastic plumbing parts.[43]

Companies for Every Facet and Aspect of Water

Beyond the kibbutz companies, two companies that had their start as government departments—TAHAL (water infrastructure) and IDE (desalination plants)—have both matured into global companies with revenues in the hundreds of millions each year. (Both are discussed at length elsewhere in this book.) Mekorot, the national water utility that is still a government-owned entity, also has a small international business serving, among other assignments, as an operator of water projects from desalination plants in Cyprus to water remediation in Mexico.

But the "What's next?" noise within the Israel water industry is coming from about two hundred young companies—representing, says Booky Oren, nearly two billion dollars in invested capital—that have all been seeded in the last ten years.[44] It seems that every month a few new tech-driven water companies come on the scene. Some are little more than ideas that can't raise funding and never get off the kitchen table, but many do. Others attract interest and funding but never develop a market outside of Israel. And a few of them are likely to grow into global companies as Netafim, Plasson, and Bermad did from their start a generation ago. The start-up culture of Israel that began in high tech migrated into traditionally low-tech industries such as energy, advertising, and textiles—and now, water.

Water is about more than what comes out of kitchen sinks or showers. It is a key part of the food supply, the energy powering homes and businesses, and the sewage that flows beneath city streets. Start-ups are popping up to address nearly every facet and aspect of water usage and supply, often coming up with solutions that save both water and energy.

With a device that looks like it belongs in a James Bond movie, Atlantium was created by an inventor who learned about laser technology during his army service. "As you look at Israel," said Rotem Arad, an Atlantium executive since 2006, "there's lots of sophistication in the light category. Israel uses lasers in medical companies. We created the world of laser-based hair removal. We use lasers in our military all the time.

So, Atlantium didn't create something out of nothing. The founders looked to see where there was a void in the application and how we may make use of it for commercial purposes."[45] Food and beverage jumped out as an opportunity.

Purifying water for food and beverage manufacturing is both essential and expensive. It can require a lot of energy. This use of energy affects the environment as well as the company's bottom line. The Atlantium product is a quartz tube in a stainless steel casing. Ultraviolet lamps came to replace the laser, but the concept remained the same. The quartz captures the UV wavelengths and directs the light toward any microbes in the water. After a few moments, the microbes are inactivated and the water is safe.

Since water is often purified by a different process that makes use of chlorine, in those cases, residues of the chemical may be found either in the food or in the factory's wastewater. Because there is no chlorine used in the Atlantium device, there are never chemical remains. Added to the savings on energy, the chemical-free method is a double win for the environment.

Atlantium's quartz and metal tube is now found in food and beverage factories in more than 150 countries. The tube is used to pasteurize water for yogurt and dairy products for companies such as Chobani and Danone, saving ninety-seven percent of the energy cost over standard pasteurization. The Atlantium cylinder is also used in water purification for Coca-Cola, Pepsi, and Schweppes soft drinks; Corona and Carlsberg beers; Nestle and Unilever beverages; and lots of other less famous brand names. In recent years, pharmaceutical companies, power plants, aquaculture, and municipal water utilities have become Atlantium customers.[46]

An Unusual Type of Public Private/Partnership

Among the highest profile people in Israel's water industry is a late-in-life convert. He still doesn't think of himself as a water guy and takes pains to distinguish himself from the "plumbers" all around him. Amir Peleg was an Israeli tech success who sold his data-analytics company to Microsoft in 2008.[47] Looking for a new opportunity, on a lark, he went to a water-industry trade show in Europe. Everywhere he walked, he saw pipes and hardware, but no software.

"Of course, companies use all kinds of data but none of them that I saw brought all of the data together," Peleg says. "This wasn't just the start of the Cleantech era, but also the start of Cloud computing and the use of Big Data. I had a hunch that there was a business there."

Israelis have an innate feel for water issues due to the climate they grew up with and their ongoing education about scarcity. When Peleg began discussing his still forming idea with tech and financial industry friends, as worthy as they all agreed it would be, they urged him to stay away. "One of them told me that he could get five exits in tech in the time that it takes to get one in water. The life cycle of water start-ups is much longer to profitability and sale than for companies in high tech," Peleg says.[48]

Nonetheless, Peleg pressed on. With leaks in municipal water systems all over the world responsible for a huge amount of water loss—and thereby energy loss—Peleg decided to become, as he now calls himself, "a high-tech plumber." By mashing together lots of historical and current data from water utilities, Peleg was able to create mathematical formulas that would be able to spot leaks, often weeks before they would be found by traditional means like soggy ground or a burst pipe. He gave the company the fanciful name TaKaDu, which sounds like it stands for something but doesn't.

The idea behind TaKaDu is to take large amounts of existing data that a utility has—water flow to a given neighborhood, conditions when a pipe is likely to burst, temperature, and thousands of other inputs—and to develop a profile. Just as credit card fraud is found by identifying anomalies in a purchase pattern, TaKaDu uses the available data and finds anomalies that are often predictive of a leak. It also helps the utility to identify the location so that digging up streets is kept to a minimum, with the expense and inconvenience that entails.[49]

TaKaDu's first customer, Hagihon, the Jerusalem water utility, wasn't just an ordinary client. When TaKaDu was still not much more than an idea, Hagihon agreed to be TaKaDu's beta site. Whatever benefit Jerusalem initially enjoyed—it is still a customer—was insignificant compared to what TaKaDu received. Hagihon's engineers gave ongoing comments on how to make the product more usable and of more value to a utility. In response, TaKaDu redesigned its software and user interface.[50]

Hagihon didn't ask for a share of TaKaDu or reduced pricing going forward. They were happy to help an Israeli water company get off the ground. "HaGihon gets the benefit of smart technology, but even more so, we help the entrepreneur get through that Death Valley period between invention and commercial rollout," says Zohar Yinon, CEO of Hagihon, of this unusual twist on public/private partnerships.[51]

TaKaDu's system is already installed in other cities in Israel, but also in England, Spain, Portugal, Australia, and other countries around the world. The utilities in these places now use the software for more than leak detection. The predictive power of the system alerts the utility's managers when to expect a surge in usage, which allows for better planning.[52]

Another Israeli alum of a Hagihon beta site is HydroSpin. Hydroelectric power is clean energy usually associated with waterfalls or mighty rivers. HydroSpin found a source so small and so obvious that—unlike TaKaDu, with its algorithms or Atlantium with its light impulses in the quartz—the first reaction might be to say: Anyone could have thought of that!

HydroSpin utilizes the water that flows through ordinary water pipes. As the water flows, it comes into contact with a very small rotating wheel. That wheel generates electrical current in the same way that a river does, just in micro-sized amounts.[53]

The HydroSpin generator addresses a coming market need. Water systems now in development will involve constant monitoring of the water for quantity and quality issues. To achieve that, the focus is moving from the control room at the initial source of the water to all kinds of meters and sensors placed alongside water pipes. The more data that is available, the more reliable and cleaner the water can be. Once the data is captured, transmitters send it to a central location for analysis. To power those meters, sensors, and transmitters, municipalities use electrical power lines or ultra-long-life batteries—both of which pose a problem.

Power lines rely on traditional energy. The more that is used, the greater the expense and environmental harm. Batteries, even of the ultra-long-life variety, burn out, and streets and sidewalks then need to be excavated from time to time with attendant cost and disruption to replace them. The mini hydro-power generators solve both problems with perpetual, renewable energy.[54]

Beyond its innovative thinking about a part of the water equation, HydroSpin represents another key element in Israel's success in water: It received funding from the incubator program in Israel's Office of the Chief Scientist (OCS). The Chief Scientist program represents the best of government engagement in industrial policy.[55]

The Office of the Chief Scientist is part of Israel's Ministry of Economy, once called the Ministry of Trade, Industry, and Labor. The incubator program was begun in 1991 soon after the influx of Soviet Jews began arriving in Israel with lots of skills and in need of jobs. Professors and scientists needed work. The government came up with a clever and efficient mechanism to help develop companies around ideas.[56]

Not wanting to be deciding where to invest, the Israeli government found a way to get others to figure that out. An incubator is a company created for the purpose of partnering with the OCS and receiving OCS funds. There is a competitive process to be selected as an incubator, with criteria that include financial strength, the skill to mentor entrepreneurs, the ability to provide labs and offices, and a history that would suggest productive relationships with potential strategic partners.

After the competitive process, a group of well-financed individuals or a company that has been selected as an incubator then serves as a screen or filter for compelling ideas that, after further development, would be good candidates to attract venture capital. The incubator evaluates many opportunities for investment and brings only the very best for consideration to OCS. Once nominated by the incubator, OCS then hires independent experts to confirm that the idea is worthy technically and is potentially the basis for a successful company.

"We don't have the desire or the in-house resources to be reviewing all of the applicants," says Yossi Smoler, the head of the incubator program in OCS. "We look to the incubator to do most of that for us. Since the incubator will have to invest its own money alongside ours, we assume they are only bringing what they believe to be investment-worthy ideas to us. We actually want to say 'yes' as much as possible."[57] Over seventy percent of the incubator-recommended entrepreneurs get OCS funding.

Entrepreneurs who get through the double-screening process receive the lesser of five hundred thousand dollars over two years from OCS, or eighty-five percent of their entire R&D expenses. The incubator has to pay for fixed overhead and puts up fifteen percent of the R&D budget. A third year of subsidy is possible, but after that, the entrepreneur and incubator have to make it work or the idea dies.

Once the entrepreneur begins selling its product or service, the Israeli government collects a three percent royalty, but only until the direct grant is repaid. There is no administrative charge or profit participation. The government owns none of the company, which is split between the entrepreneur and the incubator.[58]

Similar to the Office of the Chief Scientist incubator program is a business development concept run by Mekorot called WaTech, a word formed by combining WATer and TECHnology. WaTech is a matured form of Booky Oren's idea of helping entrepreneurs who had solutions to Mekorot's problems. Exactly as Oren had imagined the program, Mekorot has provided partial funding and a beta site to test the concept of approximately thirty different start-ups. The national water utility is often the product's first customer, also as Oren had proposed. Even better, the companies selected by WaTech get lots of attention from Mekorot's engineers, with a mandate to help the fledgling companies to think through how to best utilize their products.[59]

One recent example is MemTech. The young company is the brainchild of two veteran faculty members at the Technion. The professors, Rafi Semiat, a chemical engineer, and Morris Eisen, an organic chemist, noted that membranes used for cleaning water repelled the water that they should be attracting. By coming up with a new way of constructing the membrane, they saw there could be an enormous savings in energy costs as pushing water through the new water-attracting membrane would be significantly easier. Some believe this new membrane could be a breakthrough application in treating sewage while lowering the expense of doing so.[60]

Mekorot not only gave the company seed money to fund experiments, but more important provided more than one thousand hours of senior Mekorot engineers' time to think through optimizing the process. Mekorot will receive a small royalty on MemTech's sales, but Yossi Yaacoby, the head of WaTech, says "Mekorot didn't do this for the money. We did it so that we could have access to great technology and every year our energy use gets more efficient and our process grows smarter. Some of this comes from our own engineers, but some comes from the inventors we have encouraged."[61]

"There Is No Question as to the Trend Line"

At the most recent biennial international water show in Tel Aviv in 2013, Oded Distel, the head of the Israeli government's water-industry export promotion division, greeted fifteen thousand visitors. There were 250 official delegations from ninety-two countries coming to see nearly two hundred Israeli water companies on exhibition. Many of these companies were only a few years old; some had never made a sale.[62]

When Distel, Oren, and the others began thinking about an Israeli water export industry, Israel's water-related exports were $700 million.

They are now at $2.2 billion and Oren—still the big-idea guy—says there is no reason it can't soon be $10 billion. "That," he said, "would be good for Israel and good for the world."[63]

Distel is more sober, but no less optimistic. "In water, as with other industries," he said, "some companies will succeed, others will fail, some will do just okay. But in water, there is no question as to the trend line. Technology in water will transform utilities, municipalities, and agriculture. Israeli technology is a leader in all of these."[64]

Ilan Cohen, the former government official who helped Oren to persuade the government to set up Distel's water division, has both a historical and philosophical view. "We are in the place where the ancient Nabateans and Hebrews made sophisticated use of water. In our long exile, we lost those skills, but on our return it is as if we have had a re-awakened memory. As a young nation, we had to create new industries, and as an isolated nation, we had no choice but to think creatively or we would not have survived," Cohen said. "Now, I have no doubt we will succeed. We have become a water cluster. The world now knows that Israel has answers to their water problems."[65]

Nine

ISRAEL, JORDAN, AND THE PALESTINIANS: FINDING A REGIONAL WATER SOLUTION

It's not healthy to have a thirsty neighbor.
—Professor Eilon Adar

THE ARAB-ISRAELI DISPUTE OFTEN SEEMS like a conflict that will never end. But the growing water needs of Israel's neighbors have the potential to change the regional dynamic. For many years, water has served as a back channel for communication and opportunities for engagement between Israel and those in its region. As Middle Eastern water needs grow, self-interested necessity might well drive the regional parties closer together.

Both Jordan and the Palestinians share Israel's watershed. The three have a common destiny in jointly held aquifers and rivers, and so the three are natural partners in wanting to find common solutions. Many other states in the region who are in a different water system, but who have the same climate and terrain, are already water stressed. They will soon face unprecedented water shortages due to high rates of population growth, chronic mismanagement, poor planning, and significantly reduced rainfall. One obvious solution to their problems is cooperation with Israel.

Because Israel has taken the many steps discussed throughout this book, Israel is a water superpower. Thanks to decades of planning and sacrifice, everyone in Israel today gets all of the safe water on demand that they want—provided they are prepared to pay for it. The country benefits from wise water laws. It has a large cadre of highly qualified regulators and utility managers. And due to technological advances introduced by Israeli professors, scientists, and entrepreneurs, Israel's water security is only growing.

Israel already exports water to the Jordanians and Palestinians,[1] and often at prices less than is charged in Israel.[2] But Israel's Arab neighbors need far greater amounts to rise economically into a water-hungry, middle-class standard of living. Fortunately for them, Israel has not only more water to provide, but also training, technology, and technical assistance. With its world-class academic institutions, Israel can also be a home for professors and graduate students across the region in water-related subjects. Some now study in Israel, but many more could and should.

The Palestinians also have something of interest to the Israelis that would help to improve the quality of Israel's water. Major Palestinian cities in the West Bank sit on the Judean hills and Samarian mountains. As Palestinian homes and businesses produce sewage that often goes untreated, it migrates down the slopes via rivers that soon cross into Israel, creating pollution more easily addressed at the source.[3] Similarly, the Palestinians in Gaza dump a large volume of untreated sewage into the Mediterranean Sea every day, potentially complicating or befouling Israel's high-tech desalination plants up the coastline.[4] Likewise, Israel and the West Bank Palestinians not only share aquifers, but Jordan and Israel also share both the Jordan River and the Dead Sea. Even if Israel is capable of providing water for its country without Jordanian or Palestinian engagement, cooperation among the three serves Israel's own interests in protecting its water sources.

There is another reason to encourage water engagement among the parties. The political track may seem broken for now, but it need not always be so. Dialogue over water can be a vehicle for confidence-building measures that can lead to progress in some of the other areas of dispute. When relations improve, it is safe to wager that water will have played a role in that transformation. In any event, if the entire Arab-Israeli dispute can't be solved in one grand negotiation, it is still important to improve the quality of life as much as possible for as many people as possible in as many ways as possible.

Often, the longer a conflict continues, the more entrenched the parties become and the harder it is to resolve. With water, the long-festering conflict with the Palestinians has had precisely the opposite effect. The passage of time, and especially the advancement of technology in Israel, has provided solutions for what was once as insolvable an issue as the other points of contention today, such as final borders, refugees, security, and the status of Jerusalem currently appear to be. While it is still impossible to create new land or to return refugees to villages where cities or highways now stand, Israel has shown that it can produce new water. As Israel has transformed its water profile, so can the Palestinians, especially—and even more rapidly—if Israel is their partner in this effort.

Even without enhanced engagement, one Israeli water expert with years of experience helping find solutions to Palestinian water issues believes that Israel's water abundance may already provide peace of mind to Israel's neighbors, including those who still reject the presence of Israel in the region. "When people outside of the region look at us and the Palestinians, all they see is conflict," says Shimon Tal, a former head of the Israel Water Commission. "And there is certainly some of that. But it would be wrong to only see that. The Palestinians being our neighbor comes with a great benefit to them in water, and not just in training and access to advanced technology."

Tal points to the proximity of the Palestinians to the water-rich Israelis, and "a largely unacknowledged benefit" that comes with that proximity. "I'm not saying that the Palestinians would or should choose to be under Israeli control," he says, "but in a water-stressed region, having the chance to piggyback on Israel's growing water security, should be seen for the insurance policy it is. Whether in Gaza or the West Bank, they know that no matter how bad a drought may be or no matter what they do with their water, they know they will never go without water as long as Israel's water inventory is as deep as it is today."[5]

Water Under Israeli Rule

The 1967 Six-Day War marked the start of Israel's control of the West Bank. It was also a turning point for Palestinian access to underground water in the territory.

When the Israelis looked at the area they had just captured, they saw—especially in the West Bank—a water system reminiscent of their own before the creation of Israel's modern water grid. Like pre-State Israel, the West Bank was cantonized by region and even locality. Water from one area was not pumped to other areas where there might have been greater agricultural, economic, or household need. Jordan, which had ruled the West Bank from 1948 until the Israeli victory in 1967, drilled a few hundred wells in the territory, but they had narrow pipes and weak pumps. Little water was produced and what was generated mostly went for agricultural use in the vicinity of the well.[6]

Small pipes and poor master planning aside, the West Bank water system was in a primitive state in 1967. Water quality was inconsistent, varying by source and time of year, and often made unfit by contamination. Quantity was affected by the seasons. Gravity-based aqueducts—some from the Roman era two thousand years earlier—would transport water from local springs into towns and villages. Cisterns were found in many homes where cavities of different sizes would be filled with winter rains to provide for the family's water needs for the coming year. Even ceramic jars were used, as in biblical times, to carry water by hand or balanced on heads to bring water from nearby springs to homes and to irrigate small terraced plots.

In June 1967, only four of the West Bank's 708 cities and towns had running water.[7] In all, scarcely ten percent of the West Bank population, then about 600,000 people,[8] were connected to a modern plumbing system.[9] Similar to what Israel had done with its own water supply, it made the water from newly drilled wells in the West Bank the common property of all, while springs remained the property of their traditional owners.[10] In principle, these actions were taken for the benefit of the greater good of the Palestinians, although some have said that Israel's motivations were to obtain a portion of the Palestinians' water to supplement its own water supply.[11]

Today, about ninety-six percent of the West Bank's approximately 2.4 million Palestinians—a four-hundred-percent population increase over 1967—have running water piped to their homes.[12] Much of that water is high quality as more than half of it comes from Israel's own system.[13] "I give Israel shockingly good grades for providing water to Palestinians, especially in the last ten to fifteen years," says Alon Tal, a Ben-Gurion University professor of water management. "All but a small number [of Palestinians] get clean, safe water delivered to their homes." Tal speaks from a vantage point of someone with a long record of academic and professional engagement with Palestinians. He even ran for Knesset, Israel's parliament, in recent elections as a member of a party actively supporting the creation of a Palestinian state. "Even if Palestinians have justified complaints about the amount of water and about water pressure," he says, "it is still better quality and quantity than what is found in most of the Arab world, and even in parts of Eastern Europe."[14]

Politics or Pragmatism?

One major impediment to resolving water issues between Israel and the Palestinians is that—after many years—the Palestinian Authority (PA) has decided to make use of water as a tool to reinforce political claims against Israel, rather than working with Israel to find pragmatic solutions to Palestinian water needs. The PA is technically the governing body of the Palestinians in the West Bank and Gaza, but, since 2007, Gaza has been controlled by Hamas, a political rival of the PA. It is this inter-Palestinian political rivalry which may be a key to understanding the recent politicization of water.

The rise of Hamas, whose founding charter rejects the existence of Israel in any borders, keeps water issues from being properly addressed in Gaza by either Israel or the PA. But it also has a second, internal effect on Palestinian politics, pushing the more moderate PA to demonstrate to its public that it, too, can be confrontational with Israel.[15] Starting in 2008 and accelerating since 2010, the PA has chosen to use water as one key area of noncooperation with Israel.

The official Palestinian explanation for this makes no mention of the political rivalry. Instead, the PA says that, even if prior to 2008 Israel and the PA cooperated in the West Bank for the benefit of both the Palestinians and the Israeli settlers there, it now believes that any water or sewage project that provides any benefit to a settler community legitimizes the status quo arrangement over the territories the Palestinian Authority claims as its own.[16] Starting in 1995 when the PA was created and Israel agreed to give the Palestinians veto power over any new Israeli water project in the West Bank in exchange for a similar veto power over Palestinian ones,[17] Israel and the PA had a tacit understanding that they would approve each other's water projects. That came to an end with the more politicized Palestinian approach of recent years, a policy which unquestionably harms the well-being of the Palestinian people far more than any disruption of service for West Bank settlers—even if the goal of the PA's water policy is, more generally, to inflict harm on Israel's reputation by challenging its humaneness in connection with something so basic as water.

One example of how this standstill in approvals came to harm the interests of the Palestinians can be seen in the long-delayed Israeli approval to link up the newly constructed West Bank city of Rawabi to a waterline under the control of Israel. Rawabi is the inspiration of Palestinian business magnate Bashar Masri, and it is the first privately developed West Bank city.

The planned middle-class city Masri designed is being built in stages, with about thirty thousand inhabitants expected when construction on the land Masri controls is completed. Ultimately, though, the city may grow to be as large as 150,000 people when it is fully built some years from now. Rawabi already is a source of employment in the West Bank, and has the potential of providing thousands of new, private-sector jobs for Palestinians. As such, Rawabi is a boost to a Palestinian economy dominated by government jobs largely funded by foreign donors.

But Masri's project became a victim of Israeli-Palestinian political skirmishing and also to the Israeli insistence that the parties follow the long-agreed protocol that required them to meet to discuss the connection of the waterline to Rawabi. The Palestinian Authority refused to convene a meeting on principle if any Israeli–West Bank projects were to be on the agenda.

As this quiet conflict played out, Masri couldn't close on the apartments already built. Construction slowed, and new building stopped. Masri became ever more concerned that his project's carrying costs would cripple the future of Rawabi before the water would be turned on and the owners made their final purchase payments.[18]

An unlikely solution was achieved when Masri launched a campaign *in the Israeli media* publicizing the situation. Rawabi became a cause célèbre in Israel as a symbol of bureaucracy preserving protocol at the expense of helping Palestinians with outlooks possibly not very different from middle-class families inside Israel. Finally, after a delay of more than a year, the water connection to Rawabi—which previously would have been handled as a simple administrative matter—was opened in February 2015 on orders from Israeli Prime Minister Benjamin Netanyahu.[19]

While the Palestinian Authority ultimately won the narrow philosophical and political argument, and Israel received bad publicity in some international media, hundreds of Palestinian families were long delayed in their move to their new homes, Palestinians working on the site were laid off to protect the development company's cash flow, and Masri's venture was put at risk of bankruptcy. Ordinary Palestinians suffered for a political strategy of questionable impact.

Many Palestinian water professionals disagree with what has been labeled an antinormalization campaign—namely, the PA's policy of noncooperation with Israel on water issues—but few are prepared to speak publicly and confront the stance. One Palestinian who has worked on regional water issues with Israelis for many years said: "If we are going to solve the water problems of the Palestinian people and of the region, there is no choice but in working together. The idea that we Palestinians benefit by refusing to work with Israelis isn't smart. It actually works against us."[20]

"Fixing Palestinian water needs," says Alon Tal, the Ben-Gurion University professor, "requires a level of pragmatism the Palestinian water officials haven't been willing to show. They've politicized water and prefer to talk about rights rather than discussing solutions to the everyday water problems of their people. They prefer to blame Israel for everything than to take charge where they can and when they can," he says.[21]

In reply some Palestinians, even those in official positions, acknowledge benefits from engagement with Israel, but they quickly return the conversation to the political context. "The Palestinians learn lots of high-tech examples from the Israelis," says Almotaz Abadi, a senior official at the Palestinian Water Authority. "We visit with Israel and get studies from them regarding water and wastewater. This enhances the thinking of Palestinian managers and water engineers. But don't exaggerate this. It is a very small positive because everything else is dominated by the occupation. We can manage our own water."[22]

Gaza: Waiting for the Water to Get Worse

Gaza is in the news mostly because of the miniwars with Israel that break out every few years. But the greatest threat to the well-being of the people of Gaza might well be that they are only a few years away from a water crisis of unimaginable scope. If not addressed in time, the impending crisis will result in an environmental disaster permanently changing the quality of life there.[23]

If the complaints in the West Bank center on the water pressure and need for more water, at least there is general agreement that water pumped to homes is of a consistently high quality and, with rare exceptions, safe to drink. But Gaza, a small territory along the Mediterranean coast about forty miles from the West Bank at its closest point, is unconnected to the West Bank's water system. The two may be joined by national identity and political aspiration, but not by geology. Israelis and Palestinians alike agree: The water in Gaza is bad and growing worse.

Unlike the West Bank, which has deep aquifers and multiple sources of water, Gaza gets most of its water from a shallow reserve found as little as a few dozen feet below the generally porous soil. The shallow aquifer—the Southern Coastal Aquifer—allows for easy access for drilled or dug wells, but also for pollutants to percolate from the soil surface into the freshwater.

Professor Yousef Abu Mayla, a hydrologist at Al-Azhar University in Gaza, explains how Palestinian agriculture has contributed to Gaza's looming water emergency. "There is high unemployment in Gaza, and this led to people turning to farming," he says. "They often use inefficient irrigation techniques that not only waste water but which also permit fertilizer to get into the aquifer."[24] Gaza utilizes sixty-five percent of its available water for agriculture, in what is largely an urban area, placing a strain on an already overburdened source.[25]

Agricultural mismanagement is only part of the aquifer's degradation. "The municipality in Gaza City only provides water once or twice a week," says Professor Abu Mayla, speaking of the territory's largest city, which is home to over one-third of Gaza's estimated 1.8 million people.[26] "In response, someone in nearly every apartment building dug a well to get all of the water they needed." In all, Abu Mayla estimates, there are more than twelve thousand wells in Gaza of which only twenty-five hundred went through the permit and inspection process. This led to overpumping of the aquifer, and when the well was not dug properly, as he believes is overwhelmingly the case in Gaza, contaminants and impurities also percolated into the aquifer.[27]

Another threat to Gaza's water supply is the failure to treat sewage. Every day, about twenty-four million gallons of sewage are either stored in growing pools of human waste or dumped untreated into the Mediterranean Sea.[28] A lot of the stored waste in Gaza spills out of the holding pens. Some portion of that sewage percolates through the ground into the aquifer, further polluting Gaza's drinking water.

But the greatest problem for Gaza's water future isn't the fertilizer, pollutants, or human waste leaching into the soil. It is that the Southern Coastal Aquifer, also called the Gaza Aquifer, has fallen victim to the laws of hydraulics, a field of science that studies the properties of liquids. As freshwater is withdrawn from the aquifer at a faster pace than annual rainfall can replenish it, the delicate barrier between inland freshwater and salty seawater has begun to be breached. An overpumped coastal aquifer doesn't just empty out; it keeps rebalancing. Seawater fills in where freshwater had been. The salinity of the aquifer grows.

Seawater infiltration is accelerating. "Ninety-six percent of the water resources in Gaza are saline and in a few years everything will taste like seawater," says Fadel Kawash, a former chairman of the Palestinian Water Authority. "Almost all of the water taken from the aquifer now needs to be treated in small desalination facilities all over Gaza, but that isn't enough because the water is also polluted. Treatment removes the salt, but not the pollutants in the water people are drinking."[29]

Gaza's population has been among the fastest growing in the world. From 1967 when the Israelis arrived in Gaza until 2005 when they left, the population grew from about 350,000[30] to about 1.2 million.[31] The population in Gaza is projected to rise to over two million people by 2020.[32] Even if there were high-quality water governance and long-range planning in Gaza—which there are not—this unsustainably high growth rate would put a strain on demands for both freshwater and sewage treatment.[33]

In just the short period since June 2007, when a coup d'état brought Hamas to power, the water governance in Gaza has gone on a downward trajectory along with the quality of the local water. This ensures that, unless bold steps are taken, there will be no natural water to drink in Gaza within a few years. A United Nations study says that "irreversible damage to the aquifer" could come as early as 2020. But even if all illegal wells could be sealed and all pumping from the aquifer were to stop immediately—a nearly unimaginable event—it would still take decades for Gaza's main source of water to fully recover from its hypersaline state. Failure to act soon, though, could turn the aquifer unusable "for centuries," a UN report predicts.[34]

Many Palestinians argue that Gaza's water problems are primarily the fault of Israeli restrictions on imports and the free movement of people in and out of Gaza. They also make claims—rebutted by Israel—of damage to the water infrastructure as a result of the Israeli military operations in Gaza since 2008. But even if so, these arguments only underscore the reality that there is no logical solution to Gaza's water crisis without Israel playing a leading role.

During the period of Israel's rule of Gaza that began with the conquest of the Egyptian-held territory in 1967, Israel agreed to provide from its own resources an amount of water equal to whatever was used by the Israeli settlements and farms located there. After the Israeli withdrawal from Gaza in 2005, when the territory was turned over to the control of the Palestinian Authority, the Israelis continued providing the same relatively modest amount even though Israel was no longer using any of Gaza's water. More recently, Israel agreed to double that amount, but even if it were to be doubled again, it would be inadequate to ease the impending water crisis and societal collapse that will follow in Gaza.[35]

In the near term, the only solution to Gaza's water needs is for Israel to provide large amounts of desalinated water to Gaza from its own supplies. This creates an ideological issue for Hamas, which is opposed to any normalization of relations with Israel and, as part of that, chooses to have no business contacts with Israel. Palestinian officials, though, say they would prefer to build a desalination plant in Gaza to provide water for Gaza and then to build a pipeline across Israel for the Palestinians in the West Bank.

"Even if Hamas were able to finance, build, and manage a desalination plant," says Fadel Kawash, the former head of the Palestinian Water Authority, "it would still need to coordinate with Israel, something they swear they will not do. A desalination plant in Gaza would have to purchase additional electric power from Israel, and would likely even have to accept Israeli technical support in running the plant and in assisting with the development of Gaza's water system. The Israelis build and run desalination plants for countries all over the world and have a lot of technical expertise that we don't have."[36]

Aside from having to reverse its vow to never recognize Israel, Hamas would also have to foreswear further attacks on or infiltrations into Israel from Gaza. Before Israel is likely to permit Hamas an unfettered right to import items like cement and metal pipes that could be used either for civilian purposes (like a desalination plant or a wastewater-treatment plant) or to build weapons or military infrastructure (like rockets, missiles, and tunnels), Israel would want to be sure none of the imported products would be used to wage war against them. For now, Hamas hasn't agreed to modify its refusal to recognize Israel or to demilitarize Gaza.

The problem and its solution are clear: Gaza can be saved and more misery for the people of Gaza averted, either with Israeli desalinated water or, when conditions are right, with a Gaza desalinated water plant of its own. In the meantime, Israel could sell water to Gaza or, even better, Israel could swap its desalinated water for Gaza's sewage. Israel could treat the sewage and then use the reclaimed water for its agriculture in the western Negev adjacent to Gaza, benefitting Israel, the Palestinians, agriculture, and the environment.

But doing nothing will lead to catastrophe. The people of Gaza will be without water for drinking, washing, or agriculture. The volume of sewage in Gaza that is dumped into the Mediterranean untreated will grow. Israel will face a humanitarian crisis on its doorstep which, even if not of its making, will almost surely create political and security complications for the Israeli government. And by doing nothing, an irreversible environmental collapse may soon occur.

The Training of Trainers

Beginning in the early 1990s, Israel's and the Palestinian Authority's respective Ministries of Agriculture began designing training programs for Palestinians covering a range of subjects, which were to be held inside Israel. Nearly all of these courses had a significant connection to water. Israel had been conducting training programs in developing countries around the world since the late 1950s—and in Egypt since the early 1980s—and began doing so in the West Bank and Gaza in 1968, soon after the Israeli conquests there. This was an extension to the Palestinians of the Israeli programs already well received elsewhere.

The programs' curricula were developed in coordination with Israel's Ministry of Foreign Affairs by a division of the Israeli Ministry of Agriculture called CINADCO, an acronym for the Center for International Agriculture Development Cooperation, an organization that has provided training in nearly every developing country in the world. The intent of the Israeli-Palestinian program was to share Israel's practical experience with Palestinian water and agriculture professionals in water efficiency, brackish water, irrigation, and reuse of treated water for agriculture, among other areas.

CINADCO and the Israeli Ministry of Foreign Affairs coordinated with the Palestinian Authority to bring groups of twenty to twenty-five Palestinian agronomists, water engineers, and other professionals at a time into Israel for five- or six-day training sessions. The visitors stayed in an Israeli hotel and travelled around Israel as a group. This was done eight to twelve times a year, with the curriculum tailored to the experience and needs of each group. The seminars included classroom time, field visits, and a Friday in Jerusalem so that the Palestinians could pray at the Al-Aqsa mosque there on the Muslim holy day and also visit other Muslim sites of cultural and religious interest.

Zvi Herman headed CINADCO and its outreach program for many years. "There was no political plan or purpose in this training program," Herman says. "The goal was to do whatever was best for the Palestinian people to improve their livelihood and to see them thrive. If other benefits came from it, all the better."[37]

The purpose of the program was to "build capacity," with the participants expected to return to the West Bank or Gaza and to share what they had learned with others. "This was a training of trainers," says Herman. "Aside from the direct participants, I know that our courses touched many thousands of Palestinians' lives."

In 2010, as the politicization of Israel's water relations with the Palestinians became an everyday reality, the Palestinian minister of agriculture told the Israelis that participants could no longer take the seminar in Israel. Attending the courses sponsored by the Israeli government in Israel, Herman was informed, would be seen as a symbolic recognition and acceptance of the occupation. "With much regret," says Herman, "we cancelled the program. This wasn't a protest, but just a realization that the seminar couldn't be handled effectively when disconnected from the field experience."[38]

Around the same time that the CINADCO seminars came to an end, Israel found another vehicle for offering its water expertise, this time to both Palestinians and Jordanians. In December 1996, a Middle East cooperation organization with funding from Israel and international donors decided to create a desalination research facility in Oman ostensibly to promote the use of that technology in the region, but really as a vehicle for getting Arabs and Israelis together. It was named the Middle East Desalination Research Center, but has been called MEDRC (pronounced MED-rick) since its founding.

In 2008, MEDRC's funders expanded the group's mission to focus on helping Palestinian and Jordanian water professionals to learn about wastewater treatment and desalination of both seawater and brackish water, all of which were areas of Israeli technical expertise. Courses similar to those in the CINADCO program were designed and Israel began classes in 2010.[39] One significant difference between the two programs was that Palestinian and Jordanian experts were also part of the MEDRC faculty. Perhaps because of that difference or because of the cover provided by a regional organization, the Palestinian Authority agreed to participate. Seminars are now offered two or three times a year, down from the eight to ten times a year as was the case with CINADCO.[40]

"These MEDRC courses have had two great benefits," says Ambassador Nadav Cohen, a former Israeli water diplomat who was also involved in the design of the program and in other Israeli-Palestinian water-cooperation efforts. "First, everyone knows that the only solution to the Palestinian and Jordanian water problems will be through desalination and the reuse of sewage. This exposes their water professionals to the issues and, in only a few days, it gives them an overview of what's involved. But second, at a time when the political situation caused the Palestinian Authority's water people to refuse to meet to discuss bilateral issues, we were able to get together with them several times a year in Israel and in Jordan under the guise of a multilateral training meeting or other multilateral activities and informally discuss things we might have previously discussed in our formal meetings. As we saw, regional water cooperation can help to build or strengthen bilateral communication in other areas."[41]

Cohen points out that despite the antinormalization rhetoric coming out of the Palestinian Authority, there has never been a shortage of

Palestinians or Jordanians eager to take the courses offered by MEDRC and taught by Israeli experts.[42]

As with the CINADCO course, the MEDRC attendees are given a portfolio that is filled with information in Arabic to bring home, along with several Israeli keepsakes. The closing ceremony, as before, is filled with warmth. One of the instructors, Avi Aharoni, Mekorot's head of wastewater treatment and reuse, shared an e-mail he received a few days after the completion of one course he taught: "Hi my great teacher . . . how are you Avi, you are really a friend, teacher, and brother for me, I want to say thanks alot for you and for all, we have a good training in Israel and it was useful, I wish I could stay more of time with you. thanks thanks alot again."[43]

Bringing Jordan, Israel, and the Palestinians Together

While Israel is key to an improved water future for the Palestinians, the Kingdom of Jordan is both an example of serious effort at good water governance in a less-developed country and, especially as of late, an important partner in improving the water profile of a region all three of them share. While still facing a range of water challenges, Jordan also demonstrates the difference that long-range planning and regional integration can make in helping a country improve its water prospects.

One key element in addressing Jordan's water deficit has been to accommodate to life with Israel to a degree that no other Arab state has done. Despite Israel's own fast-growing population and a robust economy, it shares its water—fourteen billion gallons a year—with Jordan.[44] Israel does so in part because of the 1994 peace treaty between the two countries and in part because Israel sees it as wise policy to help to strengthen its eastern neighbor. More water aids the Jordanian economy and the quality of life. A stable, pro-Western Jordan along Israel's longest border serves Israel well.

While it isn't just water that has brought Jordan into a tacit federation with Israel, water has been an important driver. The countries share intelligence and security interests, and Jordan being a likely customer for some of Israel's recently discovered natural gas from the Mediterranean Sea will also help to tie their economies together. But their water cooperation has a long history.

For many years, Israel has stored water for Jordan in the Sea of Galilee because the Jordanians don't have an adequate natural water storage facility of their own. Israel and Jordan also jointly control the Dead Sea and part of the Jordan River. As important as this history has been, the integration between Jordan and Israel is quite likely to grow significantly in the near future due to a bold new project that could possibly bind the two countries— and the Palestinians, too—in a decades-long, multi-billion-dollar infrastructure venture of regional significance.

The idea is to desalinate Red Sea water and to then distribute or trade it among the three parties. Along with this, there will be an ambitious effort to undo an accelerating environmental disaster in the Dead Sea while also helping to create a new platform for regional cooperation.

The Dead Sea is misnamed. It isn't a sea; it is a lake. And it isn't dead, only inhospitable to fish and plants because of its intensely saline composition: The Dead Sea is the saltiest body of water in the world.

The Jordan River is the primary feeder of the Dead Sea. Beginning in the 1930s, as the Jewish and Arab population of the region began its dramatic growth, the Jordan River's flow was diverted, primarily for irrigation. Year after year, without the water from the Jordan River and with ongoing evaporation in a hot region, the Dead Sea began to shrink and decline. In the past fifty years, the surface area of the salty lake has shrunk by about one-third of its prior diameter and its depth has fallen by about eighty feet. It is now falling at about three-and-a-half feet a year.[45]

In one of the more ambitious water projects in the modern Middle East, Israel, Jordan, and the Palestinian Authority have come together to create a new water source for each of them while also stabilizing the

Dead Sea. Because the project can only succeed with the cooperation of all three and because it will take decades to fully reach its potential, this water-based, pathbreaking concept is a long-lasting vehicle for coexistence as well as for water. Named the Red Sea–Dead Sea Conveyance Project, it calls for a desalination plant to be built in Jordan to treat water pumped from the Red Sea near Jordan's southern port city of Aqaba, which is just across the border from Israel's southernmost city, Eilat.[46]

In general, seawater desalination removes the salt content of the water leaving behind a hyper-salty brine that is ordinarily returned to the sea. But the Red Sea has a fragile coral ecosystem and every time large-scale desalination plants have been proposed for Aqaba or Eilat, environmental concerns have been raised about the potential effect of so much brine on the coral. With the Dead Sea in need of more volume and with the shrinking lake already nearly twice as salty as the saltiest brine, transferring the brine from the Red Sea desalinated water to the Dead Sea would seem, at least at first blush, to be a clever solution all the way around, assuming environmental concerns are addressed[47] in a Phase One pilot.[48]

Without Israel's participation, it is unlikely Jordan would have built a desalination plant solely in its own territory. The Red Sea is Jordan's only access to the sea, but it is far from where Jordan needs the water. The kingdom's population and agriculture are mostly far north of Aqaba, and also at as much as a three-thousand-foot climb. To transport a high volume of Red Sea water to Amman would add an unacceptably high level of cost to the already relatively expensive process of desalting the water. By involving Israel—which has a thriving desert agriculture industry not far from the Dead Sea and which can make good use of new freshwater resources—Jordan is able to swap water with Israel and to get water to where Jordan has the greatest need.

Israel will take delivery of the Red Sea water near where the water is desalinated and, in turn, will give Jordan water from its own inventory of freshwater up north in the Sea of Galilee—which is far closer to Amman than is the Red Sea. This will save Jordan enormous pumping costs and make the project more economically attractive to banks and others providing financing.

The Palestinian have a role in this project, too, but it is not essential from a water perspective, even if it is worthwhile from a political one. By bringing the Palestinian Authority into the project, the Palestinians get a significant amount of new water for the West Bank pumped from Israel's Mediterranean desalination plants while Jordan gets valuable political cover necessary for so public an economic partnership with Israel.

"Aside from the benefits of new water and the help to the Dead Sea," says Uri Shani, a former head of the Israel Water Authority and a prime architect of the Red Sea–Dead Sea project, "the logic of the way the deal is structured is that everyone needs everyone else and everyone can only stop fulfilling their obligations at great cost to themselves. If Jordan blocks Israel's flow of our share of the water, we can stop the water being sent to Amman. Likewise, with each of us. We are all woven together. We succeed and fail together."[49]

It is especially noteworthy that the Jordanians, Israelis, and Palestinians made this deal without input from officials from other countries.[50] The World Bank, several countries, and especially the government of France, helped to provide financing for the multi-million-dollar feasibility study, but the parties jointly approached the donors, jointly worked out who would get what portion of the water, and jointly solved the many issues that a complex project like this produces at every step of the way.[51] Peace is rarely durable when imposed from the outside. As Jordan, Israel, and the Palestinians find more ways to work together without the guiding hand of others, it will assist them in turning one confidence-building exercise into another, and in aiding the development of a regional perspective on water, and perhaps much else.

Changing the Status Quo

Bethlehem University professor Alfred Abed Rabbo is an environmental chemist who does research on water science. He has what he describes as "wide experience" in working with Israeli professors and universities, particularly on the science of aquifer pollution.[52] "I'm not a politician. I'm a professor," says Abed Rabbo. "I look for solutions that help both Israelis and Palestinians. There will be no water in the West Bank in twenty years if we don't take steps together now. How are the Palestinians helped by not fixing this problem? Politics has hijacked everything."[53]

Many resent the intrusion of politics into commonsense solutions to the Middle East's water problems. But if politics is, in part, how society decides to distribute goods and services, it is impossible to avoid politics when dealing with so essential a commodity as water. The goal is to get politics, where possible, to serve the interests of conflict resolution and for water to be used by as many as possible at the lowest environmental and financial cost. The good news is that there is a lot of new thinking about how to start improving water relations, even if other parts of the conflict seem intractable for now.

One way to improve water relations is to change how the different parties conceptualize water. "If we start to think of water as a commodity and not as a symbol of national identity," says Ben-Gurion University's Eilon Adar, "we can exchange, trade, buy, or sell water in its many forms. Israel can provide desalinated water for a fee to the Palestinians in Gaza or the Palestinians there can even trade for it with the raw sewage that they aren't using and which is ruining their aquifer. Israel has a use for it. Israel can treat it and put it to use in its farms near Gaza."[54]

Almotaz Abadi, the senior Palestinian Authority water official, dreams of more than a desalinated water plant in Gaza. He'd also like to see a pipeline from Gaza built across Israel to provide the West Bank with a large amount of desalinated water. Without his referring to it as such, it would be like the National Water Carrier built to great effect by Israel in the 1950s and 60s.

Professor Adar is enthusiastic about the idea of desalinated water coming from Gaza. "As soon as possible, Gaza needs its own desalination plant and its own wastewater-treatment plants to treat all of the sewage there," says Adar. "Once the rockets and tunnels stop, there would be no reason to not encourage the Palestinians to build their own facilities in Gaza. Until then, Israel's own water and sewage treatment offers the best solution to the needs of Gaza."

Despite his support for a Gaza facility, Adar demonstrates how his approach differs from Abadi's. "If there were a desalination plant in Gaza," he says, "the opportunities get very interesting. The Palestinians could transfer desalinated water to us for use in the south near Gaza and in exchange, we could transfer more of our water from the [Western] Mountain Aquifer to the West Bank. It makes no sense for the Palestinians to incur the cost of pumping water from Gaza, which is at sea level, across Israel and uphill more than six hundred meters [about two thousand feet]. The transportation cost would add needless expense to the water in the West Bank. We built the National Water Carrier because of our isolation. No one should do that if they don't have to."

Dr. Clive Lipchin of the Arava Institute doesn't think that freshwater is even the largest concern of the West Bank. Rather, he focuses on sewage and the reclaimed water that comes from it. "In the West Bank, about two-thirds of the water is used—a lot of it inefficiently—for agriculture," he says. "The Palestinians would have enough freshwater for their homes if they were taking all of their sewage and then treating it to a level where they could reuse it for agriculture. If they were able to take even twenty percent of the freshwater now used for their crops and direct that twenty percent to their household use, the Palestinians would have more freshwater for their homes than they could use for the foreseeable future."[55]

Lipchin also believes that if the Palestinians would develop a water philosophy with reclaimed water as a central element, as Israel has done, it would change the Palestinians' view of working with settlements, some of which are not connected to a wastewater-treatment system. "If the

Palestinians saw sewage as an opportunity to power their economy, they would see hooking up sewage treatment to the nearby settlements not as recognition or acceptance of the settlements or the occupation, but as a chance to reduce pollution that threatens their aquifer and as a new source of water. Negotiating about borders or removal of settlements wouldn't need to stop. Political solutions and water solutions can be delinked."[56]

Likely every Palestinian water official or academic agrees that there is a need to build wastewater-treatment plants on the West Bank, but they are stymied—in part[57]—by a geographic problem unintentionally created by the division of administrative control of the West Bank when the Palestinian Authority was created in 1993. The Palestinian Authority was given responsibility for built-up population centers—Gaza, and what were called Area A and Area B—with the less developed parts of the West Bank—Area C—left under Israeli administrative and security control pending final resolution of the borders between a Palestinian state and Israel.

"Our sewage and polluted water are created in the parts of the West Bank that are most populated and built up," says Leila Hashweh, a Palestinian who is a graduate student at Ben-Gurion University in Israel, referring to Area A and Area B. "There is no room to build a treatment facility there." The solution, she says, is to build treatment plants in the so-called Area C, the parts of the West Bank that are found near each of the large Palestinian cities but which have few Palestinian inhabitants.[58]

While it is understandable why Israel would not want to change the Area C concept in advance of final territorial negotiations, the administrative division of the land holds everyone back, both Israelis and Palestinians alike, in moving forward on treatment of wastewater.[59] One knowledgeable American observer with sensitivity to both sides suggests the creation of a new territorial construct called Area C+ on which tertiary-level wastewater-treatment plants could be built and reclaimed water reservoirs could be sited without having to unravel the carefully constructed division of the West Bank into Areas A, B, and C. Since, presumably, all of the Area C+ locations would end up in a future Palestinian state, no one is the worse off due to this concession.

Another problem created by the Area C designation of parts of the West Bank is that—by prior agreement between Israel and the Palestinians—the Palestinian Authority has no security control over it. Palestinian thieves break into water pipes laid by Israel but located in Area C and steal water intended for communities all across the West Bank. As a result, water pressure drops and those Palestinian homes on any of the many hilly areas in the West Bank fed by one of those lines don't get all of the water intended for them.

"The Palestinian police aren't permitted to chase the water thieves if they are committing their crime in Area C," says Gidon Bromberg, the Israel codirector of EcoPeace Middle East, an environmental organization that describes itself as devoted to finding sustainable regional solutions to water and environmental issues. "At the same time, the Israeli police and army tend to think of water theft as a petty matter," not worthy of their time when they face pressures to guard against terror attacks and major crimes. "As a result," says Bromberg, "few of those breaking into waterlines are arrested, people are getting less water than they should, and the Palestinians who suffer this poor water service often grow frustrated with the Palestinian Authority, their government, for failing to deliver on a basic service." The solution, Bromberg says, proposing a unilateral concession by Israel, is for either Israel to permit Palestinian police to attempt to track down the water thieves or, if not, for Israel to police it better.[60]

As a water and environmental activist, Bromberg has long pushed for addressing water issues ahead of other regional peace concerns, but also for using water as a point of leverage in changing the dynamic between Israelis and Palestinians. One of his organization's efforts is to get Israel to make a grand gesture to the Palestinians over the natural water in an aquifer that borders both of their territories and which, by initial mutual agreement, currently favors Israel.

"Because Israel today is so rich in water, Israel can share more of its natural water with the Palestinians," Bromberg says, distinguishing between the water in the aquifer and the desalinated and reclaimed water Israel developed at great expense. "Best of all, we can share this water at a low *political* cost. Before we had a surplus, sharing more would have meant that Israeli farmers or homeowners would have had to have less. But today, Israel can share more water without having to ask for sacrifice from any constituency."

In fact, Bromberg believes, in addition to the humanitarian value, there would be great political value in giving more aquifer water to the Palestinians. "A water deal would have high political gains for the Palestinian Authority. It would show their people that cooperation with Israel delivers getting more water while rejection gets them only more of the same. For Israel, this would show the world our seriousness in resolving the conflict. We would be offering something no one expects us to offer, but in doing so it is a win for them and a win for us. It might even get people to ask of the Palestinians what steps they are prepared to take to move the peace process forward."

If broad Middle East peacemaking ideas at times seem fanciful from a political perspective, one grand idea links politics, economic development, water usage, and the environment. Clive Lipchin of the Arava Institute believes the opportunity—as far-fetched as it may sound today—goes beyond just the Israelis and the Palestinians. He seeks a regional approach, which includes Jordan, that could create a breakthrough not just in the parties' water needs, but even more by creating a sense of interdependence now largely missing. "Israel, Jordan, and a future Palestinian state," he says, "all share the region's water. What any one of them does in water or sewage will affect the other."

Lipchin proposes that the Palestinians put symbols of sovereignty aside and look to their greater self-interest. "The Palestinians do not need to develop their own energy and water grid," he says. "With a regional approach, each of Israel, Jordan, and the Palestinians has something important to contribute. Ninety percent of the Kingdom of Jordan is uninhabited, and most of it gets a lot of sun. Jordan is a logical home for a regional photovoltaic solar grid, and could provide the land for it. The Palestinians could provide the rain that falls over the mountains of the West Bank, along with the high-quality agricultural land that they have. When Gaza gets a desalination plant, it could add that to the mix, too. And Israel could provide desalinated water and the water technology it has developed, including the safe extraction of water from the aquifer."

Through this process, Lipchin says, "Israel would send water to the Kingdom of Jordan and Jordan would send clean energy to Israel. The Palestinians would protect the aquifer and would send high-quality produce to Israel and Jordan at a lower price and with less need for them to use their water resources on agriculture." This, says Lipchin, would reduce pumping from the Sea of Galilee and also help to restore the Lower Jordan River.

"Best of all," says Lipchin, "no one would be asked to give up their national identity. I'm still an Israeli and he's still a Palestinian or a Jordanian. But the doctrine of nationalism would be replaced, over time, by regionalism with all of the benefits for our economies and for peaceful coexistence."[61]

Ten

HYDRO-DIPLOMACY: ISRAEL'S USE OF WATER FOR GLOBAL ENGAGEMENT

Countries can live without an aerospace industry,
but they can't live without water.
—Oded Distel, Israeli water official

FEW COUNTRIES HAVE SUFFERED from diplomatic isolation as extreme as has Israel. In partial response, Israel has used its water know-how to assist in addressing that lonely status, often helping to develop or enhance relations with other countries. By sharing its experience and technology with others, Israel has made water an important vehicle for diplomatic and commercial engagement and at the same time has improved the water profile of countries around the world.

Even if not every nation that utilizes Israeli water expertise or technology supports Israel's interests at the UN, Israel's hydro-diplomacy has permitted it to greatly expand its international contacts. It has helped to transform Israel's relationships in the world community as more than 150 countries have welcomed an Israeli role—whether Israeli government, company, or NGO—in helping to address their water problems.

Israel's use of water for assistance and engagement in these ways has been utilized since nearly its first days as a nation. In the case of China, Israel's water expertise played a special and central role in Beijing's reversal of its long-standing diplomatic freeze with Israel. Today, China and Israel share many areas of common interest and cooperation, but few of these areas are likely to serve more in deepening relations between the two countries than Israel's assistance in addressing China's acknowledged concerns about its water.

China's resistance to diplomatic engagement with Israel began as an outgrowth of its role within the anti-Western Cold War alignment.[1] China rejected diplomatic overtures from Israel that began soon after China gained independence in 1949. The communist government in Beijing refused to have contacts with Israel on both ideological and pragmatic grounds.

Ideologically, as a leader of the communist bloc, China shunned engagement with a small state whose interests were interwoven with those of the US, a key political adversary of Beijing. But even after the US-China thaw in 1971, China continued to rebuff Israeli engagement for pragmatic reasons. First, it wanted to assure itself of a steady flow of Arab oil for its growing economy. But China was also affiliated with the Arab states at the UN and in other international venues, and did not want to risk antagonizing its Arab allies or possibly dampening their support for Chinese initiatives.

Over time, China came to realize that Israel also had something it needed.

Even though China has enormous water resources available from aquifers, lakes, and rivers, water problems plague much of the country. To cite just a few of the problems gives a sense of scope to the challenge. China's north is arid and inhospitable to farming while many farming regions elsewhere are inefficient, and often wasteful, in their use of water resources. The country's infrastructure is overburdened and loses enormous amounts of water to leaks. Sewage treatment is often inadequate. And from a legal and regulatory perspective, lax enforcement of environmental laws fails to keep water (and air) from growing yet more polluted, leading to a severe deterioration of many of China's sources of freshwater.

Despite the extraordinary disparity in size and population between the two countries, China saw in Israel a model for how it might manage its water resources.

In late 1983 and again in early 1984, China—in an episode more like a spy movie than an aid project—permitted teams of Israeli water engineers to come to China, in secret, to survey collective farms in Guangxi Province in south central China near the border with Vietnam. The Israeli recommendation was for the farms to make use of Israeli seeds thought to be a good fit with the soil and climate, and also to switch to drip irrigation. The Chinese agreed, but demanded that all markings had to be removed from the irrigation equipment and seed packaging that might indicate their Israeli origins.[2]

Three years later, again in secret, the Chinese invited another team of Israeli hydrologists and geologists to help develop an irrigation plan for the semiarid Wuwei district south of the Gobi desert. The farmers there were already utilizing all of the available local water, but using it mostly with inefficient flood irrigation. The Israelis proposed that fields be watered via drip irrigation. They also noted that the crops being grown weren't compatible with local conditions and suggested alternatives that would grow better with the available water. For good measure, the Israelis determined that there was a significant amount of underground water resources in the area that had not been tapped and suggested both how to extract it and how to transport that new water to the farmers.[3]

Soon thereafter, in early 1990, the Chinese made overtures to Israel about taking steps to develop diplomatic relations. Once again, although China had motivations other than just water, water was central to the exchange. The Chinese proposed that Israel send an irrigation and water utilization expert to Beijing and that they would send a tourism specialist to Israel. The premise was that this first public acknowledgment of Israel would not be government to government, as with an exchange of ambassadors, but would be civil society to civil society—with water being at the core of Israel's contribution to Beijing. Israel was encouraged to send a representative of the Israel Academy of Sciences to set up an office, presumably so that the Chinese could gauge both popular reaction at home and diplomatic fallout in the Arab world.

The Israeli representative, Yosi Shalhevet, had just completed a term as the chief scientist of Israel's Ministry of Agriculture and had long been associated with the Volcani Institute, a respected government research facility. Soon after arriving in China, Shalhevet began meeting with Chinese academics and others. The reaction everywhere he went was the opposite of the generally hostile news reports about Israel then being printed and broadcast in China's official media.

"Whether meeting with professors or people in agriculture," says Shalhevet, "everyone I met was excited to meet me. They had only positive impressions of Israel. When they heard I was from Israel, almost everyone would say, 'Jews, smart! Smart!' I was asked many times if I was related to Albert Einstein." Despite what they heard in the news, Shalhevet says, "Everyone seemed to admire Israel and appreciate that we were an ancient civilization like theirs. The only thing that surprised them was how small Israel was. They would joke that the entire population of Israel could check into a Chinese hotel."

A year after his arrival, Shalhevet hosted an academic conference in Beijing on irrigation that brought together ten Israeli academics and several dozen Chinese professors. "That was the first official contact between a group of Chinese and Israelis," Shalhevet says. "Within a year of that conference, I was present at the ceremony where China and Israel established diplomatic relations."[4]

While Israel has had an ambassador in China since January 1992, Huageng Pan, a Chinese citizen and a former local communist party secretary, may be Israel's greatest advocate in China; he is certainly so regarding water. After leaving politics, he started a company that manufactured energy-saving systems and water purifiers. Largely by chance, he was invited to visit Israel in 2010, and he requested the opportunity to meet with companies and professors involved with water technology there.[5] His visit convinced him that Israel had what China needed to fix its many water problems.

Since his first visit to Israel, one of many, Pan has set up a company to bring Israeli water technology to China and is now building—with local and national government financial support—an industrial park to house Israeli water companies that will give them a local presence. Culturally, the Chinese like to take time getting to know their partners, he says, and this gives the Israeli companies an opportunity for officials from all across China to get to know them. Pan predicts a robust relationship along with significant business opportunities cleaning China's lakes and rivers, remediating landfills now leaching toxins into water supplies, treating wastewater, and rethinking irrigation.

"China ranks Israel high because of the solutions that can be used to help China," Pan says through an interpreter, "but also because of the belief that Israel is a great nation and China can learn from Israel's spirit. The Israeli people have good personalities and traits that Chinese people think they can be influenced by. No matter where we go in China, everyone says very positive things about our association with Israel. It is never difficult explaining why Israeli water solutions should be used."[6]

Pan is not alone in believing that China can benefit from Israel's advances in water. In May 2013, Israeli Prime Minister Benjamin Netanyahu and his delegation arrived in Beijing's Tiananmen Square to meet with Chinese Premier Li Keqiang. A member of Netanyahu's delegation describes their arrival: "I can still remember when China refused to recognize Israel," he says. And yet upon their arrival, the delegation found a greeting that showed how clearly the relationship had changed.

"All of Tiananmen Square was filled with Chinese and Israeli flags," the Israeli official says. "It was very emotional. We were brought into the Great People's Hall for our meeting, and the two delegations sat across from each other. We were addressed respectfully and as equals."

Before the Israeli delegation had left for China, Netanyahu and his senior staff had agreed on opening comments in which he would suggest that Israel, "as a junior partner," might have something valuable to offer in helping China with water management. But Netanyahu never got a chance to make that short speech because his Chinese host made it for him. "Premier Li began the meeting by warmly greeting us," says an Israeli participant at the opening session, "and then said that China is aware that Israel knows how to manage water and that Israel has great water technology. He concluded his remarks by saying that there were many places in China that suffer from water concerns and that he hoped there would be something the two nations could do together."

Delighted that the two leaders were on the same wavelength, Netanyahu proposed picking a small Chinese city and having an Israeli consortium redo its entire water infrastructure. The implication of the offer was that if the project was a success, it would be rolled out to other cities. In reply, Premier Li designated one of his government ministers present to assist in picking the small city, evoking laughter from the Israeli delegation when he suggested the Chinese government official focus on a city of about one million people. Explaining the laughter, Netanyahu is reported to have said, "Mr. Premier, we have no city in all of Israel with a population of one million people. To us, one million people isn't a small city."

In late November 2014, a joint Israeli and Chinese selection committee announced that Shouguang, a city of a little more than one million people located three hundred miles southeast of Beijing in Shandong Province, would be the first test city in the China-Israel water relationship.[7] Population aside, the city and its surrounding area have a diversified range of water challenges that make it a logical choice. The project will address water purification and treatment of wastewater as well as efficient irrigation for the many farms surrounding the city. There will even be the need for specialized water treatment for factories and a paper mill adjacent to Shouguang. A consortium of fifteen to twenty Israeli companies using Israeli technology will help to rethink and reengineer the city's water usage.

"I don't want to get ahead of ourselves," says one senior Israeli official closely involved with the project, "but if we perform well here, we will have the opportunity to help rebuild the water systems of cities all over China. This will not only bring significant revenues to Israeli companies. It can also lead to deeper China-Israel relations for a long time to come." As he points out, "there are a lot of Chinese cities."

Israel to Iran's Rescue

If Iran is in the news because of its nuclear program, the greatest threat to the country's well-being isn't economic sanctions or the Islamic Sunni-Shiite schism. Rather, the greatest threat to Iran may be that the country is running out of water. The problem is so severe that social unrest, economic dislocation, even out-migration can all be imagined. One government advisor recently predicted, as reported in *Al-Monitor*, that as many as fifty million Iranians—seventy percent of Iran's population—may be forced to leave their homes because of a lack of water.

Water problems are a proxy for bad governance, and Iran has water problems galore. Underground resources have been overpumped beyond what can be naturally recharged by rain, and, on the present course, many aquifers will soon be unusable. Iranian agriculture is among the most wasteful in the world. Most countries use about seventy percent of their water for agriculture; Iran uses over ninety percent.[8] Even so, Iran is already not self-sufficient in food, a trend that is projected to get worse.[9]

The Islamic Republic's climate is mostly arid and semiarid, which, by definition, means it only gets modest rainfall. More than half of the country's wells are believed to have been dug illegally and many of them, possibly most, are now polluted.[10] More than two-thirds of all industrial facilities fail to treat their wastewater, and manufacturers, even of chemical products, generally dump their waste into Iran's waterways.[11] Iran discharges more than sixty percent of its sewage untreated, polluting groundwater, rivers, and lakes.[12] Climate change is only likely to exacerbate the overall poor water outlook.

A visitor to Iran looking at each of these problems and knowing that Israel has largely overcome all of them might conclude that the Islamic Republic would be wise to overcome its antagonism to Israel and invite Israelis to Iran to help manage its water sector. As fanciful—and nearly impossible—as that sounds, it is exactly what Iran's ruler, the shah, did beginning slowly in 1960 and with a rush after 1962. Israeli hydrologists, water engineers, planners, and others became so numerous and so enmeshed in Iran's water exploration and infrastructure that the majority of the water projects in Iran from 1962 until the 1979 Islamic Revolution were managed by Israelis.[13] Geopolitically, for Israel the alliance with Iran served to counterbalance the hostility of the Arab states while lessening Israel's regional isolation—at least for as long as the cooperative relationship continued.

Though not as dramatic as in the hit film *Argo*, the head of the Israeli water team, Professor Arie Issar, left Iran on the next-to-last direct flight from Tehran to Tel Aviv in 1979 shortly before the shah was deposed. He described scenes of descending chaos on the capital's streets as his car made its way to the airport. That would be the last of many trips that began for him in 1962 as part of a humanitarian project trying urgently to repair an Iranian water conveyance system from antiquity.[14]

Ancient Persia had a sophisticated, gravity-based water system used for irrigation utilizing vertical shafts called *qanats* that were dug on a slight decline from an underground water source to the fields where the water would be needed. In 1962, Qazvin Province, nearly one hundred miles northwest of Tehran, suffered a major earthquake. More than twenty thousand Iranians died, three hundred villages were in ruins, and the network of water tunnels first dug more than twenty-seven hundred years earlier was destroyed.[15] Qazvin was home to a vast agricultural valley that provided fruits and vegetables to Tehran and beyond. After the earthquake, the farmers were without essential water.[16]

Meanwhile, the shah had already been quietly cultivating a relationship with Israel. Iran believed itself vulnerable to mischief by some of the Arab states and saw Israel as a valuable counterforce.[17] The shah was also impressed by Israel's scientific advances in agriculture, water, and, ironically, in nuclear power. In 1960, he asked the Food and Agriculture Organization (FAO) at the UN to provide water experts to help advise Iran, and, with his consent, three Israeli technicians were sent. When the Qazvin earthquake struck, the shah already knew of Israel's sophistication in water planning and exploration.

On an emergency basis, Israel was invited to send water engineers to Qazvin to see if the *qanats* could be rehabilitated. A close inspection revealed that they were damaged beyond cost-effective repair. In any event, what might have been ideal for irrigation in the time of ancient Persia was no longer optimal in an era of modern agriculture. The Israelis successfully urged government officials and the farmers to abandon the collapsed *qanats* and to permit them to drill deep wells of the kind Israel itself was digging back home.[18] The Iran-Israel water relationship quickly blossomed.

Soon after well drilling in Qazvin began, Israeli water engineers received a positive reply from their Iranian government hosts to a proposal that they also be permitted to teach local farmers how to increase their yields while using less water in the process. Their interactions with Iranian farmers broadened to include advice on which crops to plant and how to market them. The majority of the local population in the greater Qazvin area came into contact with the Israeli engineers, none of whom masked their nationality or religion.[19]

Shmuel Aberbach was one of the UN's Israeli FAO experts who happened to be in Iran at the shah's invitation when the earthquake struck. A geologist and groundwater expert, he made his way to Qazvin shortly after the quake and helped put together the plan for where and how to drill the region's new wells. Over the following seventeen years, he made dozens of trips to nearly every part of the country, and got to know (and often train) Iranian hydrologists. In all that time, he says he did not experience a single anti-Israel or anti-Jewish incident except for an offhand comment from an Iranian communist who hated all of the Soviet Union's cold war rivals. Aberbach also never heard of any anti-Israel slurs reported by any of the dozens of Israelis he knew in the country except for some chants at an Iran-Israel soccer match in Tehran. Decades since his last visit to Iran in 1978, he still has close friendships with Iranians he met from his work, many of whom live in exile.[20]

Another Israeli who spent time in Iran, Dr. Moshe Gablinger, a Cornell University–educated Israeli engineer, had similar reflections about his relationships with his Iranian hosts. Although he didn't develop any lifelong relationships, he also only had friendly, cordial interactions. "We never socialized with them in their homes, but there were warm relationships," he says. "To meet an Iranian hydrologist in a restaurant for dinner wasn't an unusual experience."[21]

Professor Issar, the Israeli in charge of all water exploration and drilling operations in Iran, remembers being taken to remote corners of the country and being introduced to local residents by Iranian hydrologists who would travel with him. "They would say that I came from Israel to share our knowledge with them," he says. "I was always welcomed and invited to a special meal which they rushed to prepare. The only problem I would have was that I had to sit on the carpet on the floor and eat the roasted lamb and rice without a knife and fork."[22]

The caliber of the Iranian water professionals was not generally high. "Despite all of its oil, Iran was a poor country then and its education system didn't properly prepare its water professionals," says Dr. Gablinger. "The people assigned to me were very nice personally, but quite backward and unsophisticated technologically."[23] Professor Issar instituted programs training hydrologists and technicians, and offered classes in geology, hydrology, and chemistry.[24] Shmuel Aberbach taught advanced mathematics to Iranian hydrologists and geologists to produce predictive models on how much water was remaining in aquifers.

Iran's welcoming attitude toward its Israeli guests included some touches unimaginable today. The shopkeepers in Qazvin learned Hebrew to better engage with their new customers. Dr. Gablinger remembers most of his local interactions with the merchants being in Hebrew. Moreover, by the mid- to late-1960s, so many Israelis had arrived in Qazvin with their families that a local building was converted into a school in which Hebrew was the language of instruction for the sixty Israeli children who studied there under Israeli teachers. Even more remarkable, the shah came to visit Arie Issar and his team in Qazvin not long after Israel's crushing defeat of three Arab armies in the June 1967 Six-Day War—signaling his approval of the work of the Israelis in Iran.

The shah also encouraged delegations of Israelis in other specialties to visit and he sent Iranian officers and scientists to Israel. Some Iranian water professionals stayed in Israel for extended periods to learn advanced Israeli techniques. Commercial and political ties between the countries became deep and wide.

"The only part of Iranian society we could not penetrate was the religious establishment," says Uri Lubrani, Israel's ambassador to Iran from 1973 until shortly before the shah was deposed. "We were welcomed everywhere else. Everyone in Iran is either very religious or from a home that was. Even the Iranian communists then knew Islamic ritual. No one used religious differences to keep Israelis away except for the clerics. We tried hard, but they wanted no part of us. Arafat [the head of the Palestine Liberation Organization (PLO)] had cleverly cultivated [Ayatollah] Khomeini in exile, and Khomeini had made it clear to the religious administration to have no contacts with us."[25]

Notwithstanding the views of Iran's clerical sector, the initial success of the Israeli engagement in Qazvin was extended to many other provinces and regions. An Israeli government-owned water-engineering company, TAHAL (an acronym for the Hebrew words "Water Planning for Israel"), was asked to supervise the construction of water and sewer systems in such major Iranian cities as Isfahan and Bandar Abbas, and to create household and irrigation water systems for whole regions, like Hamdan and Kermanshah. When Mashhad, Iran's second-largest city, needed a system developed to distribute cooking gas for homes citywide, the Iranian government turned to TAHAL to do that, too.[26]

Other Israeli-government companies in related water areas were also invited into Iran. Mekorot, the Israeli national water utility, was asked to drill for water throughout Iran as it had done in Israel, and to run a large project on the Iranian part of the Caspian Sea, among other assignments.[27] Solel Boneh, also an Israeli government-owned entity which handled major construction projects back in Israel, was hired to build dams throughout Iran and infrastructure in Iran's cities.

Around 1968, IDE, the Israeli government company created to brainstorm desalination ideas, developed a breakthrough energy-saving process, and it was eager to test the concept in real-world settings.[28] At about that same time, the Iranian Air Force wanted to secure safe, clean water for its bases. Professor Arie Issar remembers that the Israeli military attaché in Tehran, Colonel Yaakov Nimrodi, saw this as an opportunity to use Israel's water expertise to deepen the Iranian-Israeli military connection. Nimrodi arranged for IDE to be invited into Iran. Over the following decade, IDE installed thirty-six small desalination units in Iranian Air Force facilities and nineteen others around the country.[29]

In 2007, nearly forty years after IDE began installing desalination systems in Iran and long after the Islamic Republic had severed all ties with Israel, Fredi Lokiec, an IDE senior executive, was at a trade show in Europe when he was quietly approached by an Iranian engineer. The Iranian told him that several of those aging Israeli desalination units were still in use and that Iranian technicians had attempted to reverse engineer one of them so that they could start building copies of the Israeli unit from scratch inside of Iran. He said that they got the copied plants to work, but never as well as the ones built by Israel.[30]

After the 1979 revolution in Iran, Khomeini and his supporters held mass trials of Iranian government officials and others thought to have been the shah's supporters. People of the Bahá'í faith were also at risk and widely persecuted. Both Arie Issar and Shmuel Aberbach had Iranian friends and colleagues in Iran's water industry, a few of whom were Bahá'í but mostly Muslims, who fled the country in the first days of the revolution and who still live in exile. Tragically, both Israelis also knew Iranian water officials who were executed for still unknown crimes. With Israeli water experts expelled from the country and many Iranian water professionals exiled or executed, Iran's water industry was dealt a lasting blow that sowed the seeds of Iran's coming water calamity.

Helping More Than One Hundred Less-Developed Countries

Beginning in the late 1950s, Israel began sharing its irrigation and water techniques with less-developed countries, initially with a special emphasis on Africa. While there was the potential of diplomatic and trade benefits in deepening ties with these former colonies and less-developed countries, at least initially, the focus was mostly altruistic and an outgrowth of Israel's Zionist philosophy.[31]

In his 1902 political tract-cum-novel *Altneuland [Old New Land],* Theodor Herzl, the founding prophet of Zionism, has the novel's protagonist announce that following the establishment of a Jewish national home, Jews needed to help the people of Africa, because they had "a problem, in all its horror that only a Jew can fathom." After the national "restoration of the Jews," Herzl's character says, the next task would be "to pave the way for the [national] restoration of the [African] blacks."[32]

Israel's founding generation of leaders were guided ideologically by Herzl and took this admonition of his to heart.[33] Although Israel in the 1950s was itself still a developing country, David Ben-Gurion, a socialist, a Zionist, and Israel's first prime minister, would say that even if the country had little to share, the emerging nations of the post–World War II era needed to share hardships as well as fellowship and ideas.[34]

In 1958, Golda Meir, then Israel's foreign minister, created a department in her ministry whose mission was to help developing countries—with a focus on Africa—overcome problems with water, irrigation, agriculture, education, and the status of women. The department was called MASHAV, a Hebrew acronym loosely translated as Center for International Cooperation. The Israeli emissaries sent abroad were mostly farmers and engineers, many of them veterans of the British Army's World War II Jewish Brigade from which they had gained experience working in economically underdeveloped British colonies during the Mandate era.

In its early years, MASHAV (and Israel) was warmly embraced by African states as well as countries in Asia and South America. Israel made clear that MASHAV would not provide cash gifts or grants as did (and do) US and European aid programs. "We called our efforts development cooperation and never called it aid," says Ambassador Yehuda Avner. "We were there to help by teaching and training, but not by providing financial assistance." With Meir's focus on Africa, in only a few years, MASHAV programs were widely present there, and hundreds of Israeli specialists, in water and other areas, were living and teaching on the continent. When Meir became Israel's prime minister in 1969, she saw to it that MASHAV and its African program continued to get the support it needed.

In what proved to be a "heartbreaking episode for Israel and Golda [Meir]," says Avner, in the aftermath of the 1973 Yom Kippur War, every sub-Saharan African nation broke diplomatic relations at the urging of the Arab League and the Organization of Islamic Countries. The Israeli specialists there under the auspices of MASHAV were all expelled. "This was bad for Israel, but was a personal trauma for Golda," says Avner. "She had been messianic about her African program, and it all came to naught."[35] Meir and Israel aside, it was also an unfortunate turn for the many Africans aided by programs whose improved water, irrigation, and food projects were abruptly terminated.

In the 1980s, some African countries expressed interest in renewing ties. Ethiopia restored relations in 1989, and the rest of sub-Saharan

Africa—eager to see a return of Israeli water professionals and other experts—did so in 1993 after the first Oslo Agreement between Israel and the Palestinians was signed. Today, Israel provides training in water management, irrigation, and other areas for specialists from more than one hundred less-developed countries, twenty-nine of them in Africa. Many of the programs take place in Israel and others are conducted in the host country. In all, water and irrigation training still account for as much as forty percent of all MASHAV outreach programs.[36]

Ambassador Haim Divon served as head of MASHAV for eleven years, and traveled widely in the developing world. "The US and Europe might have more sophisticated programs for less-developed countries than Israel does," Divon says, "but Israel, with all of our on-the-spot improvisation, may actually be a better inspiration. We can show these countries what we have achieved in water and other fields in fifty years. Israel is a successful model and one that is actually within reach of these countries. When they see a system like America's, they don't think they will ever be able to achieve that."

Since MASHAV's beginning, more than 270,000 people from 130 countries have participated in MASHAV programs.[37] "That is a large number," says Divon, "but a drop in the bucket when compared to the billions of people who live without adequate food, water, and secure futures. There's still so much more to do."[38]

Serving the Poorest of the Poor

While Israeli water technology and products are found in more than 150 countries around the world, one Israeli water company, TAHAL, has specialized in water and related activities in the developing world. TAHAL has had a reach and impact in those nations without equal among Israeli companies; its work has improved the quality of life for hundreds of millions of the world's poorest people.[39]

TAHAL was started by the government of Israel in the early 1950s to plan and design complex water projects in Israel. But by the close of that decade, nearly all of the major projects to be developed in Israel had been designed. Eager to avoid layoffs, the government company sent a senior executive on a trip to recently decolonialized nations to see if the work the company had been doing in Israel might be relevant in those other countries.[40] It was.

By the mid-1960s, TAHAL's five-hundred-person organization had staff located throughout Africa, Asia, and South America involved in developing the water supply and sewage disposal for major cities and in designing irrigation plans for large farming districts throughout the developing world.

Within a few years, TAHAL's presence in some countries became so significant that the company became a quasi-governmental department of the client country. With important future benefits, TAHAL was recruited by several African governments to advise them on their infrastructure projects, even when TAHAL itself was not involved in the initiative. Perhaps because of this central, even intimate, role, when Israeli government aid workers were expelled from Africa after the Yom Kippur War, TAHAL saw no material disruption of its work in Africa.[41]

Serving as an advisor on these many projects did more than make TAHAL invaluable to its host countries. It also led to an expansion of TAHAL's expertise, helping the company to grow in later years from exclusively doing low-fee design and engineering work to taking on bigger-ticket construction, procurement, and management projects. It also planted the seeds for a significant expansion of TAHAL's mission. TAHAL went from working only in water, sewage, and irrigation to pursuing, among other projects, opportunities in environmental and agribusiness, and even, recently, in natural gas.

In the 1990s, following Margaret Thatcher's UK privatization model, the Israeli government began selling Israeli government companies including its national airline, all of the state-owned banks, and the country's telephone monopoly. In 1996, TAHAL was sold, too, leading to significant growth for the now-independent company. The company today has twelve hundred employees who generate more than $250 million in annual revenue largely focused on dozens of projects in thirty less-developed countries, but it has worked in dozens of others. It still also succeeds in winning the design competitions for about seventy percent of all water-related projects in Israel, but doesn't compete to build water projects elsewhere in the developed world.[42]

"TAHAL's success," says Saar Bracha, the company's CEO, "comes from our connection with Israel. Everyone knows that Israel made the desert bloom by finding water in unusual ways. Everyone knows we used special irrigation technology to get better yields with less water. But when people think of water in Israel, even if they don't know TAHAL's exact role, TAHAL gets some of the credit—and justifiably so—for many of those great achievements. When we call on a country with water needs, they already know us, even if we've never worked together before, because they know what Israel has achieved."[43]

Sometimes TAHAL's Israel connection has worked against the company's business outreach. Despite India's many water problems, India—like China—refused to have any trade or diplomatic relations with Israel for the first four decades after the two states gained independence less than a year apart in the late 1940s. India was one of the three leaders of the anti-Western Non-Aligned Movement. In deference to India's Arab partners in the movement (and also out of concern that ties with Israel would inflame India's large Muslim minority), Israel's diplomatic overtures were rebuffed by India.

Although India today still often votes against Israel's interests at the UN, the two countries are now bound up in trade and defense matters that have led to genuinely warm relations.[44] But until the late 1980s when trade began between them, India rejected contact of any kind with Israel. Diplomatic relations were established in 1992 and TAHAL won its first assignment in India in 1994.[45]

Since winning that first assignment, TAHAL has had a large role in the modernization of significant parts of India's water system. Whether developing master plans for states like Rajasthan and Gujarat, or creating and implementing irrigation infrastructure for Andhra Pradesh and Tamil Nadu, or designing and building the sewage system of Assam, TAHAL has been hired for each of these following a competitive process run by the World Bank in partnership with the Indian state involved.[46]

Recently, TAHAL showed potential in another line of business: running part of a major Indian water utility.[47] Although a municipal utility is ordinarily a government function, India, like other governments around the world, has experimented with the public-private partnership (PPP) model in roads and bridges, among other parts of its infrastructure. A decision was made in India in 2012 to see if a major water utility might be better run by a company with a for-profit discipline.

Partnering with the Jerusalem water utility and an Indian infrastructure company, TAHAL and the others entered into a novel arrangement in 2013 with the water utility of Delhi, India's capital city. Delhi was designed by the British, late in its long colonial rule of the Indian subcontinent, to be a city of 800,000. While there have been many additions to roads, electricity, and plumbing since the British departed in 1947, the city's population today of more than sixteen million people has overwhelmed its infrastructure.

Throughout Delhi, water is often only available when purchased from a water truck. In this joint venture, TAHAL and its partners have been brought into two neighborhoods—the chic, upscale Vasant Vihar and the crowded, poor Mehrauli, with a combined population of more than one million people—to rethink, rebuild, and run the water utility. If the experiment is a success, the project will be extended to other parts of Delhi and, perhaps, to other parts of India. TAHAL would likely also add running developing countries' municipal water utilities to its list of business areas.[48]

Years before TAHAL was permitted into India, the company had begun working in the region. Beginning in 1975, TAHAL spent more than twenty-five years working in the lower Himalayas in Nepal. The focus of TAHAL's work there was in Bhairahawa and in Lumbini, Buddha's birthplace, where the company had been called upon to develop underground water resources and to build an irrigation system so that the impoverished local farmers could make use of that newly found water.

"You can hardly imagine how terrible the conditions were," says Dr. Moshe Gablinger, the now retired TAHAL executive. "There were no roads and no electricity. It was the most primitive living conditions for us. One morning a member of our team woke up and saw a snake in his room." Gablinger had many international postings with TAHAL—among them ten South American and five African countries, and was head of operations in Ghana—but says he never experienced anything quite like the first years in Nepal. "By the time we left in around 2000," he says, "we could say that it was one of the most successful irrigation projects ever done. It changed the lives of all of the many poor people there."

Gablinger believes that the willingness of TAHAL and other Israelis to go to what he calls "miserable locations" is a key reason for Israel's success in water-development projects around the world. No matter how underdeveloped the destination, Israelis would sign up for the task. Coming from a comfortable middle-class life, earning an Ivy League Ph.D. in engineering in the US, Gablinger thinks there were three reasons that made TAHAL people, and Israelis in general, do what they did.

"First," he says, "there were many people of great talent," and he modestly excludes himself, "who just needed bigger challenges than Israel could offer them. They wanted to call Israel home, but didn't want to be limited to the projects that Israel had to offer, especially after most of the water infrastructure was built and water was flowing to the desert."

The second motivation was their pride in Israel, Zionism, and Jewish tradition. "Everywhere we went," he says, "we wanted people to know we were from Israel and that we were Jews. We wanted our work to remind everyone that this is what Israel does. And because we were from Israel, we demanded of ourselves that wherever we went, we'd always behave at our best. If TAHAL benefited from the Israel connection, we also wanted to be sure that Israel's reputation benefited from our work."

Gablinger says that of equal importance to the other two explanations was altruism, even if with a twist. "We would travel to these places with no hint of modern conveniences, and we would cry for the people there," he says. "It was an honor that we could help poor people and poor nations, and improve the quality of their lives. It was almost like a commandment from the Bible, this feeling we had of wanting to help people all over the world."[49]

An NGO Runs African Water Systems Remotely from Tel Aviv

The Israeli government and Israeli companies hardly have a monopoly on Israel's water innovation in developing countries. Sivan Ya'ari is a case in point. Ya'ari is a force of nature, the kind of person who achieves results in large part through the felt aura of her willpower. What this diminutive thirty-something Israeli wills now is for her still young NGO—known as Innovation: Africa, or i:A—to utilize solar power and Israeli technology to help bring clean water and electric power to people living in small, often remote, villages in Africa.

Israeli born, her parents moved to France when she was a child. Ya'ari came to the US for college. To pay her tuition, she sought work and was introduced to one of the owners of Jordache, the fashion jeans company, who hired her to do factory inspections at manufacturing facilities in Africa. Like others before her, Ya'ari was amazed by the stark difference between the life she knew in Israel, Europe, and the US with what she saw on her trips to Africa, most especially the lack of necessities as basic as clean water and electricity.

The experience moved her, and without knowing where it would lead, she sought a graduate degree in international energy management. That led to a summer job with the UN in remote parts of Senegal where she saw water pumps either broken or villagers unable to afford the cost of diesel to run them. "They had pumps right there," Ya'ari says, "but because they couldn't use them, they ended up digging bore holes a few kilometers away to get filthy water they had to carry back to their villages." She describes it as a heart-wrenching experience.[50]

Returning to New York and her studies, before her planned return to Israel, she started Innovation: Africa. Her idea was for the organization to install solar-powered electricity for lightbulbs and vaccine refrigerators in clinics as well as solar-powered water pumps. In January 2009, following a bit of modest fund-raising by Ya'ari, Uganda's Putti village became the site of the organization's first solar-powered water project. Others have followed, and there are now projects in seven African countries.

Though i:A's Ugandan efforts still only reach a tiny percentage of the country's millions of villagers who lack clean water, the NGO has caught the eye of some national leaders in Africa. Uganda's Prime Minister Ruhakana Rugunda told an interviewer that he appreciates the work of Ya'ari and Innovation: Africa and that "it is a good expression of the cooperation between Israel and Uganda."[51]

Ya'ari—a mother of three young children and the founder of a national chain of New York–style nail salons that employs more than 150 workers throughout Israel—plans for a significant expansion of the water project that now, she laments, "only" reaches tens of thousands.

Finding water has not been as hard as Ya'ari had anticipated. "It turns out," she says, "that there is a lot of underground water in Africa. You just have to know where to look for it. The bigger problem facing African water-assistance programs is that as soon as the aid professionals leave the villages, the systems begin to break down and the people are no better off than they had been." Ya'ari decided to overcome that with smart technology, all run remotely from Israel.[52]

Innovation: Africa has been able to create a system that seems to be impervious to breakdown, vandalism, or theft—the problems faced by water systems installed by other aid organizations. The concept is deceptively simple. Quality underground water is located, and a rented diesel-powered drill is brought in to reach it. A water pump is inserted into the shaft. Solar panels, sized in order to produce the electricity needed to operate the pump, are installed and connected to the pump. The pump draws water out of the aquifer and deposits it into an adjacent water tower built for the purpose. When the water in the tower is needed, it is propelled by gravity to destinations all around the village.

Since having adequate food is a concern of villagers nearly as significant as access to clean water, the i:A waterlines are also connected to a drip-irrigation system installed at the same time as the solar panels. The local villagers only need to plant seeds adjacent to the drippers. Everything else, except for the harvest, is handled thousands of miles away in Israel.

Meir Ya'acoby, i:A's technology chief, is an electrical engineer who has worked at the Israeli R&D centers of major US technology companies. He now splits his time between his tech start-up and his work with i:A. Using simple components, he created a device that allows each African water system to be monitored and managed from the i:A office in Tel Aviv. By utilizing whatever data or cell phone wireless service that is available in the African village ("They may not have shoes, but the adults have cell phones," Ya'ari says.), frequent messages are sent updating key information such as how much water is in the tank, or if there is a problem with the pump, the drip-irrigation equipment, or any of the solar panels.

What Ya'acoby adds to the transmissions from Africa is a constant feed of information on local weather conditions. If it is going to be hotter than usual or if it will be cloudy for a few days blocking the solar rays from producing electricity, he knows to have the pump put more water in the tank as a precaution. If there will be rain, he knows, depending on which crop is being grown and where it is in its growing cycle, when to stop the drip irrigation from providing water and also when to restart it. If a mechanical problem develops anywhere in the system, he knows about it as quickly as a few minutes later and the system can automatically contact a local engineer with detailed information to fix whatever has gone awry. Ya'acoby can automate every part of the system so it is infinitely scalable, he says.[53]

The drip-irrigation systems are also having an unexpected effect beyond providing more food in the village and relief from hunger. "The people eat what they need and they sell the extra at the market," says Ya'ari, citing what happened in Putti village in Uganda as a representative case study. "With the extra money that they made from the sale of drip-irrigated produce, they bought chickens and have developed a poultry farm. This has improved both their nutrition and also now given them some financial security. The water in these systems can become an economic development tool. And," she says, "with the success of the village tied to the water system working properly, everyone makes sure it is protected."

There are now dozens of villages around Africa that already have i:A systems, and the results are quickly felt. "Once you begin providing water," Ya'ari says, "the children become clean because they aren't filling jerry cans with muddy water and because they can wash. They also stay healthy, because a large number of the children had been getting sick from drinking unclean water." Another change is how they spend their days. "The children, especially the girls, had been walking two to three hours a day fetching water," she says. "They would come back exhausted and filthy. Now, with water being pumped, they can go to school. They have no obligation to be fetching water at all. For them, water is for drinking and bathing."[54]

Eleven

NO ONE IS IMMUNE: CALIFORNIA AND THE BURDEN OF AFFLUENCE

People should realize we are in a new era. The idea of your nice little green lawn getting watered every day; those days are past.
—California Governor Jerry Brown

If we want things to stay as they are, then things will have to change.
—Giuseppe Tomasi di Lampedusa, *The Leopard*

UNTIL RECENTLY, NEARLY ALL of Israel's overseas water projects took place in economically distressed or underdeveloped locations. As Israeli inventors and entrepreneurs developed new water technologies, though, Israel's global water footprint grew to include affluent and water-rich countries and communities. Many Israeli water-technology companies now do business around the world, providing water solutions in wealthy countries, even in those with abundant water resources.

These Israeli innovations touch almost every part of the water profile, including the manufacture of freshwater from desalinated seawater, sophisticated approaches to irrigation, advanced concepts in sewage treatment, cutting-edge metering and leak-detection systems, and an array of new efficient technologies that enable the whole gamut of water systems to work with greater reliability and energy efficiency. Whether saving water or reducing cost and energy consumption to run those water systems, increasingly Israel has made a difference in these well-off places as it has in less-developed ones.

But Israel has something more to offer the affluent world than products and services. It also has the benefit of its water experience to share, most especially the role of innovation and taking on issues before they become problems.

Ordinarily, utilities and farmers in affluent societies are slow to innovate in how they manage and utilize their water. Likewise, even when government leaders saw the first glimmers of water problems years ago, few were eager to push for meaningful changes of the kind that would preempt water crises. For as long as everything seemed to work more or less as it had before, no one felt impelled to call for water to be priced at something close to its real cost or even for modest sacrifices by voters and business leaders. If citizens and industry had no interest in seeking changes—and why would they know that they should be doing so?—elected officials had no interest in imposing them either.

Although Western countries all enjoy sophisticated water infrastructure and safe, running water available on demand, most of these countries have taken their water abundance for granted. Many have sleepwalked into counterproductive legal and regulatory structures, while their citizens, agricultural sectors, and industries have carelessly adopted wasteful—even destructive—consumption patterns.

A changed approach to water is needed nearly everywhere, and affluent countries have the means to effect that change in a way that less-developed countries do not. Their farmers and consumers can better afford to pay the real cost of water. Their infrastructure is more sophisticated. Their governance is likely more flexible. Their citizenry is more conditioned to activism. Most significantly, their larger educated publics can be made to understand why change is needed. Without thinking of water as the finite resource that it is, and without accepting some modest restraints now, consumers and farmers may soon—and perhaps abruptly—find severe restrictions placed on their water usage. Affluence provides an advantage but not immunity from the coming global water crisis.

Just as the Israel of the 1950s was a model for many poor countries in their use of water, the Israel of today—an affluent society with advanced water governance and policies—can serve as an example for rich countries and regions. This includes those who have not yet felt the sting of water problems and who have time to get ahead of the concerns still over the horizon. That Israel sits in a water-limited region as it has seen its population and economy grow, while also creating a water surplus, offers hope everywhere that with enough time, investment, and changed attitudes, everyone can have comfortable water futures.

Without planning for the worst, though, even water abundance can be reversed without much notice. Brazil is a case in point.

A Brazilian Nightmare

After experiencing a generation of rapid economic growth, Brazil lifted tens of millions out of poverty and into comfortable middle-class lives while the country also became the great economic success story of South America and one of the most attractive countries for investors. Brazil is also the home of the Amazon River—by far, the largest flowing river in the world, an irony, perhaps, for those living on the other side of the nation in São Paulo, Brazil's economic capital. Due to a combination of drought and bad water policies, São Paulo is now suffering a water shortage that would have once seemed irreconcilable with the city's office towers, luxury hotels, and elite precincts.

Ten percent of Brazil's two hundred million people live in the city of São Paulo and twenty percent of the country lives in the state also named São Paulo, where the city is located. São Paulo accounts for a third of Brazil's GDP and forty percent of its industrial production.[1] But São

Paulo's current water crisis raises the question of what future this world financial center and its surrounding area will have if its water problems aren't soon fixed.

In early 2015, water in São Paulo's main reservoir dropped to about five percent of capacity. With most of the region's power coming from hydroelectricity, there wasn't enough volume to run the electricity-generating turbines. Rolling blackouts affected São Paulo and the eight states around it, affecting the health, safety, and economy of tens of millions.[2] Because of the lack of power to run the pumps and also because of the reduced volume, tap water stopped running in many homes for days at a time, and then only intermittently when it was available. Restaurants couldn't serve meals on plates as there wasn't enough water to wash them. People lacked water to shower—and even to flush toilets. Water fears became the central part of everyday life in a metropolis not unlike major urban centers in the US.

In response to the water shortage, area residents began drilling into the local aquifer without permits, imperiling underground resources long into the future with contamination. Thieves broke into the city's already leaky waterlines and stole what water they could, further depleting supplies. People set up makeshift storage systems in their homes and apartments to capture what little rain might fall or that might flow through their taps, solving one problem while creating another: Mosquitos began breeding in some storage tanks, and cases of dengue fever were reported.[3]

If civil society did not break down, then widespread anger at the government and the local water utility was widely expressed for failing to anticipate problems, fix infrastructure, and establish proper water governance before and during the eight-year drought. Grim thoughts about a mass exodus from São Paulo were heard, and those who had already departed the area were given the label "water refugees."

California Turns to Israel

Nothing comparable to São Paulo's water shortage—and crisis—has been felt in a major city in the US. But after only a few years of drought in the American west, it is no longer impossible to imagine a future of water restrictions and a change in the way of life in the US, even if one not so dire—at least, not yet—as has befallen Brazil's most populous city and state.

Like Brazil and São Paulo, though, only a few years ago, the US's most populous state would never have thought itself vulnerable to water shortages. When thinking of California, one conjures up images of amply watered lawns, glistening and frequently washed cars, and backyard swimming pools adjacent to bubbling Jacuzzi baths.

Parts of California, most especially Los Angeles, the state's biggest city, may have been located adjacent to a desert, but the glamorous, larger-than-life lifestyle there implied abundance in everything, including water. As for the state's farmers—growers of some of the world's finest produce and as everywhere, the largest users of water[4]—they had long been subject to restrictions from using river water in order to protect endangered fish species and were given assurances that they could make use of available sources of water such as other surface water resources and nearby aquifers.[5]

Today, without much warning and with many of those aquifers that had been offered in reserve now pumped so low that they have gone dry, California is planning for a very different future. The once water-assured state recently announced a series of measures limiting water usage to protect itself from what may be an ever worsening drought. After a dramatic state-imposed cut in household use, farmers announced limitations on their own water use to fend off larger, mandatory ones. Likewise, resorts all over the state enhanced their use of recycled water for golf courses, fountains, and theme park attractions. And California also did what countries with water problems around the world have been doing for decades: It turned to Israel for partnership and help.

Following years of conversation about deepening economic ties with Israel,[6] the California drought gave urgency to the talks. In January 2014, Governor Jerry Brown declared a drought state of emergency for California. "We can't make it rain," he said, announcing the declaration, "but we can be much better prepared for the terrible consequences that California's drought now threatens." To make explicit how dire conditions were, he spoke of the risk of "dramatically less water" for farmers and for everyday life. The announcement left little to the imagination in reminding Californians that the state needed to be ready for fires, the risk of which was made worse by the drought, and even for a shortage of drinking water.[7]

Although long in the works, less than two months later, Brown welcomed Israeli Prime Minister Benjamin Netanyahu to a ceremony in Silicon Valley to sign a Memorandum of Understanding (MOU) to begin the process of creating a "strategic partnership for enhanced economic relations."[8] Behind that stated goal was a focus on shared innovation with Israel, with water at the top of the list.

When it was his turn at the podium, Brown gave several reasons for wanting Israel to be the first country with which California was establishing ties to jointly address global concerns. But he began his remarks with a recitation of California's water problems.

"We are in the midst of a megadrought," Brown said, "and very seriously this just brings home to us how important it is to manage our water—efficiently and wisely." He then acknowledged the needs, but also the challenges. "We have a long way to go in water conservation, recycling, use of desalinization, managing [water] both above ground and underground. And Israel has demonstrated how efficient a country can be and there, I think, is a great opportunity for collaboration."[9]

In reply, Netanyahu provocatively posed a question: "Israel doesn't have a water problem," he said, "and you may ask how is that possible?" He laid out that Israel's rainfall was now half of what it had been when the State of Israel was founded, that the population had grown tenfold, and that GDP—usually the surest measure of water consumption—had grown seventyfold. The explanation, Netanyahu said, was found in Israel's recycling of sewage for agriculture, drip irrigation, prevention of leaks, and desalination. "We don't have a water problem," Netanyahu repeated, "and California doesn't need to have a water problem. By cooperating together, I think that we can overcome this. We've proven it. It's not something we talk about as a possibility. It's a concrete result."[10]

Unlike other government-to-government agreements that depend on the follow-up activities of officials, this MOU was different. It called for a structure to foster cooperation between Israeli and California businesses and universities. It also urged California's cities to find ways to work with Israel in addressing their water—and—other problems.

Soon after the MOU was signed, top schools in the University of California (UC) system, including UCLA and Berkeley, among others, began finding projects under the MOU on which they could cooperate with Israeli professors and universities. Two California cities created task forces to utilize Israeli solutions to specific problems, including Los Angeles, which sought Israeli help in reversing problems with a polluted aquifer that supplies parts of the city with its drinking water.[11]

Kish Rajan is a top aide to Governor Brown. He heads the governor's business and economic development efforts in a state that would be the eighth-largest economy in the world if it were a country. "There is hardly a sector in California that isn't touched by water issues," he says. "Agriculture is the most obvious, but not the only one." The state's agricultural sector is a $70 billion industry. "With a growing middle class around the world and a demand for high-quality food for them, California has the opportunity to increase sales of produce, but only if we can increase yield and manage our water."

The export of produce is, in a sense, the export of water, even if it is the fruit, vegetable, or nut that overseas consumers believe they are buying. When water is seen as a free or inexhaustible resource, farmers have little incentive to factor in the cost of water in the cost of the export. But if the supply of water becomes scarce, it becomes a policy question as to whether growing produce for export makes sense—at least until new water resources can be identified and assured. By extracting what may be largely nonrenewable water from an already overdrawn aquifer and thereby risking a state, region, or country's water future for the short-term benefit of a sale obviously makes no sense. Even so, it is what is happening in parts of California, and around the world.

Different from agriculture, which is an export-driven business, is tourism, another major California industry. It is a $100 billion contributor to the state's economy. "A key part of our tourism industry," says Rajan, "is the California lifestyle, a lifestyle important to both visitors and the people who live here. That lifestyle includes recreation, golf, swimming, and lush landscapes. All of that requires a lot of water."

The value of the Israel relationship in helping California maintain its way of life and to grow is significant. "Israel has a system that allows them to manage water intelligently. They know how to charge for water and they understand how to use technology to create efficiency in the use of water. All of these have allowed Israel to be as successful as they are. This is a new era for us in California, but they have been doing this for a long time. And we have a lot that Israel can teach us in each of these areas," Rajan says.[12]

One Israeli input into California's water revival that Rajan might have mentioned is the desalination plant in Carlsbad, near San Diego, now being built by IDE Technologies, the Israeli desalination company, after ten years of delays caused by litigation and permitting. The state-of-the-art Carlsbad plant uses all of the newest technology that IDE has developed at desalination facilities that IDE built and manages in Israel and around the world. When up and running, Carlsbad will be the largest desalination plant in the Western Hemisphere, producing fifty million gallons of water a day or, at current consumption levels, enough water for three hundred thousand Californians. If not enough to reverse the effects of a punishing drought and the accumulated deficit of many years of robust water use throughout the state, it is a significant step forward, and one of many needed to reverse a problem not going to go away by itself.

Forty States Expected to Suffer Water Shortages

What California is now doing on its own and in partnership with Israel to bend its water curve is widely relevant to other US states, and in other affluent countries around the world. Even if California's problems have received a lot of publicity, it isn't the only state challenged by a reduction in its water supply. Much of Texas has suffered recently from a drought so severe that even after the drought restrictions had been declared partially lifted, an official with the US National Weather Service said, "I'm not sure it's ever going to get back to where it was."[13] Dozens of communities in Texas still face chronic water shortages for agriculture and, in some places, even for everyday use.[14]

As with California, the Texas economic boom was threatened by its own punishing drought, climaxing in 2011, but with lingering effects to this day. In just 2011, the state lost nearly $12 billion from agricultural and other failures. More than three hundred million trees died. Dozens of communities were within weeks of running out of water, and nearly fifty have yet to recover.[15]

In October 2013, Texas Governor Rick Perry travelled to Israel to meet with Israeli officials and to develop partnerships akin to California's that could help Texas through the worst of its water troubles. As important, while still in Israel, Perry said that it was necessary for Texas to not only limit the impact of the drought then beginning to recede, but to also prepare for the inevitable next one.[16] As Perry's comment implied, the economic losses from the drought, and the misery it caused for millions of Texans, would have been reduced or even avoided with planning and earlier spending on infrastructure.

Drought alone of the kind that California and Texas have endured isn't the only threat to the prosperity and livelihood of those who work the land in many US states. Farmers in all of the eight states that sit above the massive High Plains Aquifer—South Dakota, Wyoming, Nebraska, Colorado, Oklahoma, Kansas, Texas, and New Mexico—have a common concern. They know that the days of unbridled pumping of water out of the aquifer and the irrigation of crops like alfalfa, corn, and wheat with that seemingly inexhaustible supply are over. If these crops are to be grown as they have been for decades, it is only a matter of time before the water will run out. What took millennia to fill the giant, nearly continental, aquifer has already been significantly depleted in decades.

The answer to these farmers' needs can be found in developing crop varieties that grow well with less water, and in technology that can help them to grow more with less water and less aquifer-polluting fertilizer. As in Israel, the High Plains farmers need to make every drop count, and to find ways of reusing every drop so that the remaining water in the aquifer can last long into the future.

Agriculture and industry in other states are also suffering. Nevada, New Mexico, and Arizona continue to struggle with chronic water shortages. The Colorado River, which has sustained much of the West since river-diversion projects began about one hundred years ago, has reached historic low flow thanks to drought, overpumping, and inadequate planning. Idaho, Oregon, and Washington State also now feel the effects of drought-induced hardships.[17]

Beyond those fifteen states briefly profiled or mentioned here, the US Government Accountability Office (GAO) reports that forty of the nation's fifty state water managers—including those from these fifteen states—expect freshwater shortages in their states within the next ten years. Although the projected geographic scope and intensity of need differ from state to state, what is consistent is a lack of adequate data, an uncertainty regarding how best to respond to this expected upcoming water challenge, and the lack of a coherent national plan.[18]

Also impeding progress in all of these affluent states is a regulatory structure that gives many a say in how water should be managed, but with few compulsory guidelines. With a vast bureaucratic structure numbering in the thousands of entities, coordinating decision making becomes nearly impossible. Every water board, water authority, water council, and related government entity not only is inclined to preserve its prerogatives, as is true of bureaucracies everywhere, but the existence of so many governing bodies also confounds the ability to plan ahead, to fix current problems that go beyond the geographic footprint of that entity's power, and to raise needed funds for infrastructure whether by bonds or taxes.

A Way Forward

In all of this bad news now coming from California, from the forty states expecting water problems, and also from the affluent countries across the world beginning to feel the gap between their water needs and water supplies, there is also the good news that every one of the water problems has a solution. And in each case, Israel can serve as a model based on its own experiences, its own trial and error, its own failures, and ultimately its own solutions.

Consider just a few examples of challenges Israel faced—and overcame—in the use of its water and what that experience could mean to the US in enhancing water resources for the benefit of farmers, consumers, and the environment:

Israel makes use of drip irrigation on seventy-five percent of its irrigated fields, reducing water use and increasing yield. In those places in the US where rainwater is inadequate, use of drip irrigation should be accelerated. In just the last few years, California's use of drip irrigation has soared, and with very positive effect.[19] More can be done there, and everywhere else. Government policy should assure that the cost of drip-irrigation equipment is never an impediment to adoption. Special tax treatment of the equipment's depreciation and government loan programs can accelerate the ability of farmers to start saving water—and to make flood irrigation an anomaly rather than the national norm.

Treated sewage is now seen as a precious resource by Israel's farmers and eighty-five percent of that country's sewage is collected for treatment and reuse. The current goal is to bring that up to ninety percent within the next five years. Nationally, the US treats the vast majority of its sewage, but then dumps nearly all of it in lakes, rivers, and oceans. Less than eight percent of US sewage is reused.[20] With smart, but long overdue infrastructure spending, farmers can lower the demand for freshwater by using this resource, while treated sewage from population centers like Los Angeles can be more widely used in the watering of golf courses and public lawns.

Israel has built five desalination plants along its relatively short coast—including one that is the largest and most energy-efficient in the world. In addition, it has built several desalination facilities to make use of otherwise worthless brackish water. Israel built its coastal desalination plants in less time than it took California to overcome legal issues just in building the Carlsbad plant. While much of the US is too far inland to make use of desalinated water, California, Washington, Oregon, and other coastal states can address their water shortages, in part, by building more of these plants. State and federal governments can establish criteria to expedite permits so that litigation is limited and necessary manufactured freshwater can be added to the natural freshwater supply.

The management of Israel's water is undergirded by a legal and regulatory structure to maximize the use of its water. Although untangling the state and federal regulation and management of water in the US seems almost impossibly complex, few actions that state legislatures and Congress could take would have so positive a systemic effect on the future of water in the US.

Israel encourages innovation and technology in every part of its water system. The far wealthier and more technologically advanced US scientific community could be partnering with state and local governments, farmers, and utilities to introduce water- and energy-saving techniques.

And finally, Israel has created and continually reinforces a culture that accepts, and even admires, water conservation as a part of everyday life, while its people are able to live comfortable middle-class lives. No comparable mass mind-set exists in the US to impress the point that conservation doesn't require suffering. In a water-limited world, saving water today allows for comfortable living tomorrow.

When Israel decided it wanted to enhance its already adequate national water infrastructure, it made the decision to charge the real price for water to finance it. In so doing, national consumption dropped quickly and significantly, in farm and city. Charging higher prices for water wasn't primarily done to conserve water, but it was a positive outcome leading to long-term changes in behavior. Rational—and universal—water fees in California, and everywhere else, will inevitably lead to better agriculture policy, more innovation, greater citizen engagement, and an end to reckless water usage.

If the best outcome from the coming global water crisis now rolling over the US and other affluent countries would be for everyone to become inspired by what Israel has done in long-range planning, conservation, pricing, and water use, the worst outcome would be if bureaucratic inertia, the self-interest of powerful groups, timid elected officials, risk-averse utilities, and a distracted public cause some or all of the known water challenges to be pushed aside to be addressed another day.

While Israel has invented many of the solutions that have changed the world of water, what sets the country apart isn't the technology—all of which is known and available to all—but rather the *extent* to which it has adopted these techniques. Throughout Israel, one can find posters exhorting citizens and visitors to make every drop count. It is that mindset that may be the most important solution of all for a water-starved world.

Water shortages in the US may seem more shocking or unreal, perhaps, than in faraway places like Brazil—or certainly than in less-developed countries—because the US has long seen itself as the land of plenty. Resources have always seemed as inexhaustible as sunlight or air, and abundance a part of the American birthright. But with water scarcity already here and growing, the choices are to prepare for it before the worst of it hits, as Israel did beginning in the 1930s, or to prepare to suffer unknown, but dire, consequences. Today's actions will mostly be felt in ten years or longer. With the global water crisis accelerating, there is little time to lose.

PART IV

HOW ISRAEL DID IT

Twelve

GUIDING PHILOSOPHY

If you don't know where you are going, any road will take you there.
—Lewis Carroll, *Alice in Wonderland*

In about ten years, beginning shortly after the new century, Israel went from scarcity of water and fear of drought to abundance and independence from climate conditions. This dramatic change was made possible by the seventy years that preceded it in which a cadre of often brilliant engineers, scientists, and policy makers developed Israel's water-related expertise, technology, and infrastructure. A pragmatic water philosophy also evolved from these leaders and visionaries to guide the way for those who would follow.

Israel is both an old nation and a young country, with each part of this split personality contributing to the country's stability and success. Its millennia-old traditions and attachment to the Land of Israel provide its sense of rootedness and strength in a harsh, unforgiving terrain and region. The modern identity of the country—a new state that prizes fresh ideas and disruptive thinking—gives it a restlessness that results in experimentation and comfort with change. Shimon Peres, Israel's former president, told an interviewer while still in office that "the greatest Jewish contribution to the world has been dissatisfaction" which, he said "is bad for the country's leaders, but very good for science and progress."[1]

The "restless Israel" has developed a broad acceptance of core principles for how best to manage its water. This consensus provides one explanation as to why Israel is the water success it is today. Israel has enjoyed many triumphs; military, technological, sociological, and economic among them. But its success in water is no less remarkable. "Out of seven billion people in the world today," says Hebrew University professor of hydrology Haim Gvirtzman, "only about one billion have truly safe, always available, high-quality water. Most of these one billion are in humid areas like North America or Europe. What is remarkable is that Israel—which is in an arid region—has both safe water and reliable systems. This is harder to achieve than you would imagine."

Not everything that Israel has done in water—even its successes—should logically be copied everywhere. Some countries have great amounts of natural water or rain and don't need to desalinate water or build reservoirs to capture water from a fleeting rainy season. Some countries are just too poor to afford all of the elements that a modern, water-focused country of accelerating affluence like Israel has put in place. But if some of the techniques, infrastructure, or technology Israel now employs aren't a fit for everyone, Israel's philosophy in water governance may be.

Like every country, Israel has a unique national identity. But Israel's culture and history don't have to be adopted to make all or part of Israel's water philosophy the underpinning of a country or state's water worldview. Israel's ideas can be adapted to a variety of economic and social settings.

The following dozen elements are separately and together a key to understanding Israel's philosophy (and success) in water.

"The Water Belongs to the Nation"

Even in their dynamic, free-market country, Israelis believe public ownership and government management of water achieve the best outcomes for all. Beginning in the 1930s and codified by the forward-looking 1959 Israel Water Law, all of the water found in Israel is common property. This has permitted Israel to plan for the greatest water needs of the society as a whole while taking into account all of the available resources.

Consistent with their willingness to surrender control of water to the government, Israelis across the economic and political spectrum are bewildered by a (chaotic) free-market approach to water. "Israel prioritizes water based upon highest, best use," says Haim Gvirstman, the Hebrew University professor who also did advanced research at Stanford University in California. "In the US, it's a free-for-all, and with what result? To cite just one example, there isn't enough water for cities in California and Arizona and, at the same time, large amounts of water are being used for agriculture for crops that could be grown elsewhere. It is illogical to squander water with flood irrigation on farms while people downstream in Los Angeles and other cities have water restrictions. In Israel, the water belongs to the nation and we decide how to best utilize it for the greatest good."[2]

Embedded in this facet of Israeli water philosophy is the importance of centralized planning, even within a country that generally otherwise prizes free-market solutions. "We govern the whole cycle of water, from the first drop until its final use," says Professor Uri Shani, former head of the Israel Water Authority, the independent government body that coordinates and authorizes how much water is produced and from where as well as to whom that water goes and at what price. "Water management is completely centralized. Every pump and borehole, every allocation of water, has to receive a permit. Planning for and allocating every drop of our water has been the key to our success."[3]

Cheap Water Is Expensive

Consumers have been trained to think that the lower the price, the more satisfied they should be. Usually, that's true, because the price paid for goods or services reflects its real cost with profit added. Buyer and seller both benefit. Water is the international exception to this bedrock principle of economics. Around the world, subsidies are the norm as nearly no one pays the real cost for what they use, especially not for the water needed to grow the food they eat. In Israel, those who use the water cover the full cost, with not a penny of government subsidy.

"The real cost of water," says Gilad Fernandes, an economist and a senior official at the Israel Water Authority, "includes the development of the water resource, the infrastructure that has to be built to transport it, the testing and treatment of the water so that it is safe to drink, the pumping of it to the home so that it is always available, and the removal and treatment of sewage so there is no danger to rivers or aquifers." Although a few other countries also use real-cost pricing, in much of the world, it is common for consumers to pay little more than the pumping cost to their homes or a flat monthly fee, if even that.[4]

The most important reason for setting water and sewage fees at their real price is to let market forces work. Real pricing encourages consumers to use all of the water they need, but not more. Israel has shown real pricing to be the most effective conservation tool of all.

With market forces at work in Israel, farmers, who are the largest users of water everywhere, make decisions about which crops to grow by taking account of the real cost of growing them. To avoid unnecessary expense and waste, farmers are incentivized to use the best available technology to save water. As the market for water-saving ideas developed in Israel in response to the cost of water to all users, more entrepreneurs began directing their capital and ideas at developing ways to reduce water use even more. A virtuous circle of saving water and technological innovation was born that would grow far larger if water was priced at its real cost everywhere.

For consumers familiar with hyper-expensive bottled water, full pricing of freshwater may sound like a large burden and give reason to fear having to pay for their household water at those stratospheric levels. In the real world, however, the full price of water is much less than most people imagine. But even at very low prices, real pricing has a deep and lasting effect on water consumption.

Israel had long subsidized water. In recent years, it abandoned doing so in favor of full pricing. Yet, for most homes, prices come out to less than a penny a gallon, or less than twenty-five cents for an ordinary shower. For very heavy household users, the price rises to about a penny-and-a-half per gallon, a pricing mechanism which keeps prices lower for light users of water. Despite this very small cost, by ending across-the-board subsidies, Israel transformed the country's demand for water. Usage in Israel dropped by nearly twenty percent.[5]

Israeli officials explaining the pricing system for water often contrast it with sunshine, something properly seen as free and endless. Full pricing for water helps to transform it from a free good that can be used without any restraint to a commodity with limits.

Use Water to Unify the Country

There aren't many benefits to Israel's small size, but in water management, it has been a blessing. Water has been moved to where it is needed by Mekorot, the country's national water utility since before the state was created. "Competition can lower costs," says Ronen Wolfman, now one of the heads of the China-Israel water company Hutchison Water and a supporter of business competition generally, "but multiple utilities would have led to duplication and either reduced service or higher costs. Instead, Mekorot can operate in the public's interest everywhere."[6]

Israel's water is blended from many sources and no one receives preferential treatment in water quality or access to more quantity. Everyone prepared to pay gets as much water as they want. Poor people have their water paid for by the same social welfare agencies that help the indigent with rent, food, and medical expenses—but someone pays for every drop of water.

Likewise, regardless of where they live, all consumers pay the same price for their water. If you live adjacent to a well with water a short distance from your home or on a mountain that requires expensive pumping to get the water to you, the price is the same. While this nationally blended price means that not everyone pays their personal real cost for the water they use, it does result in everyone paying the same price and everyone having a common, unifying stake in conservation and innovation.

Water serves to unify the country in another way: It is a source of pride for Israelis that their country has overcome all kinds of impediments to have its region's most sophisticated water system, an infrastructure at least equal to the wealthiest countries in the world—most of which are in water-rich locales.

Yossi Shmaya, an executive who manages one of the largest water facilities for Mekorot, spoke for many when he said, "I'm so proud of our achievements in water. Everyone you meet in Israel who works in water will say the same. This isn't just a job. It is a national mission. You can compare Israel not just to our neighbors but to all over the world, and no one has done what we have done."[7]

Or as Ori Yogev, a senior government official and former water entrepreneur, said, "Our conquest in water was like winning a second war of independence."[8]

In much of the world, water is a source of disunity. Israel has found a way to use it as a source of national cohesion.

Regulators, Not Politicians

Decisions about water would seem to be ripe for politics. Politicians routinely decide who gets what in society. In theory, at least, if politicians misallocate a resource, they will be voted out of office and the newly elected officials will fix the problem. But Israel regards water as too important to be left to the whims of politicians.

Because of electoral reality, elected officials around the world are usually reluctant to spend more on water. The benefits of new water infrastructure aren't felt until long into the future, probably after the elected official is gone from office, or at least from that office. Raising taxes or issuing bonds now to pay for expensive water infrastructure for which credit will be given to a successor makes little political sense. Public money can be spent instead on more visible projects like parks, schools, and hospitals that are likely to translate into more immediate popular support. Fees collected for water and sewage can even be used to cover shortfalls in unrelated parts of the government budget.

But as with other citizen-driven political movements—most notably the environmental one—the priorities of politicians will follow the interests of broad public engagement. Until that day comes, though, politicians are most likely to either ignore water issues or to favor allocation of water to political supporters with an interest in water.

To keep special interest groups and friends of elected officials from getting favored treatment and to keep spending on infrastructure, technology, and innovation high, Israel decided to keep politics and politicians out of water decision making. A technocratic regulatory structure with centralized authority—the Israel Water Authority—was created and power was taken away from an array of government ministries supervising water.

Mirroring the national regulatory structure, each city and town has a nonpolitical local water utility. The mayor appoints the board of directors, but each candidate is required to have specific skills before he or she can be nominated. The goal at the local level is the same: to take politics and politicians out of water decision making.[9]

Create a Water-Respecting Culture

Across Israel, there are signs posted that remind consumers to conserve water. The role of every citizen in saving water begins to be taught in the earliest grades, and the principle becomes ingrained. The public may not enjoy water restrictions or flow reducers on showers, but they understand why they are needed.

A positive effect of Israel's water-respecting culture is that it creates a partnership between government and the governed. When periodic droughts strike, the public understands what is expected of them. Compliance with water-reduction efforts is widely honored.

This ongoing conservation training serves the country's water interests in more than just times of water shortages. The we're-all-in-it-together mind-set helps to unlock citizen activism in finding new ways to save water and to not waste it.

Water in Israel is in the domain of the government. But water innovation has become the domain of any Israeli person, company, or organization willing to innovate in a market always eager for new thinking. This approach deepened the sense of partnership on water between government and citizen.

All of the Above

Consider what Israel does in pursuit of clean, safe, available-anytime water:

- Pumps and purifies natural water from its aquifers, wells, rivers, and the Sea of Galilee.
- Desalinates seawater.
- Drills deep wells to get brackish water.
- Develops seeds that thrive on salty water.
- Treats nearly all of its sewage to a high level of purity and reuses it on crops.
- Captures and reuses rainwater.
- Discourages landscaping of parks or homes that consume freshwater.

- Seeds rain clouds to enhance rainfall.
- Demands all appliances (especially toilets) be hyper water efficient.
- Replaces infrastructure before leaks begin and promptly fixes leaks when they appear.
- Educates schoolchildren as to the value of water conservation.
- Prices water to encourage efficiency.
- Gives financial incentives for technologies that save water.
- Experiments with ideas to reduce evaporation.
- Transformed its agriculture to grow water-efficient crops.
- Uses drip irrigation for most of its agriculture.

What makes this list so extraordinary isn't just its depth and comprehensiveness. Rather it is that it represents the Israeli conviction that there is no single answer to Israel's water worries. Obviously, some techniques produce, or save, more water than others. But even with the easy surplus that desalination has brought, Israel's water professionals have effectively pursued an "All of the Above" approach that consciously integrates all possible sources of water and all possible technologies for conservation.

"Intentionally building a national system with redundancy and overlapping supplies of water," says Shimon Tal, a recent head of Israel's Water Commission, "is expensive and requires expertise in many areas. It means our bureaucracy must be larger than a more narrowly focused approach would be. On the other hand, it is also liberating because we know that Israel will have high-quality water anytime people want it, that our economy and agriculture can grow, that we can welcome new immigrants and millions of tourists, and that we mostly don't need to share the worries about shortages of water that people all over the world and certainly all over our region have. Any one part of our program can go down—a desalination plant to a war or an aquifer to drought—and no one will have their water shut off."[10]

Use Water Fees for Water

The creation of municipal water utility corporations throughout Israel took away local control of water management from city hall and gave it to a local technocratic board focused solely on water and sewage management. Under the new governance structure, one hundred percent of water and sewage fees are used for their intended purpose: assuring a local and national water system of excellence.

With predictable and sufficient revenue now in hand, the Israel Water Authority has been able to finance two key goals. First, it wanted adequate funds to fix leaky local infrastructure while also building a national system to transport desalinated water from the Mediterranean Sea. Second, the Israel Water Authority wanted the municipal utilities to use more technology and innovation.[11]

More is now spent on both of these, and with the hoped for results. Billions of gallons have been added back to the country's supply each year by stopping leaks. Utilities have become high-tech beta sites for credible ideas. When a new technology succeeds in one city's water system, it is quickly adopted by other Israeli water utilities. With this approach, every city in Israel has the potential to become a laboratory of water innovation—and Israeli water entrepreneurs know they will have access to a new kind of public-private partnership.

In most places, utilities aren't known for rapid innovation or as first adopters in technology. Israeli utilities have gone from risk-averse traditionalists to centers of innovation. Israeli consumers also know that their water fees are being used to make sure that Israel will continue to stay ahead of its water needs.

Innovation Wanted

By popular agreement, Israel's water sector is centrally controlled, with pricing, allocation, and planning in the hands of a technocratic government authority. Even so, government policy is to encourage privately driven innovation and public-private partnerships.

In just the past decade or so, more than two hundred water-based start-ups have begun operations in Israel, equaling about ten percent of water start-ups worldwide in this period.[12] Most of them are based on innovations to existing technology, even if a few are breakthrough ideas tied to entirely new approaches to use of water or sewage. Just as the Israeli government encouraged the creation of kibbutz industries in the 1960s and 70s, it also spurred the creation of these many ventures through a national mind-set welcoming to new ideas and by not stigmatizing failures. As before, this generation of start-ups is also often provided with financial incentives.

A special kind of holding company—the incubator—exists in Israel to find innovations and to apply for government support in getting the entrepreneurial idea off the ground.[13] Similarly, Mekorot, the government-owned national water utility, not only provides funding to start-ups with promising ideas, but also assists with their product development by donating what could be as much as thousands of hours of senior executive time—for a product owned by a private company.[14] Israel's municipal utilities also receive government subsidies to serve in a similar role for real-world testing of new ideas. These utilities provide the support of their in-house engineers at no charge and are encouraged to share the best ideas with other municipal utilities.[15]

By seeing the private sector as a partner in developing Israel's sophisticated water economy, national water regulators prevent some of the impediments common in other government or bureaucratic settings around the world. Turf wars are very rare in water innovation, as is the "not invented here" syndrome that can cause the sabotage of fresh ideas in water and other bureaucracies elsewhere.

Israel also makes use of its own government assets when it is deemed best and turns to the private sector for solutions when lower price or greater innovation can be had. Despite Mekorot's sophistication in desalination, the government chose a private consortium to build most of

Israel's seawater desalination facilities as it believed the private sector could deliver water at a lower price. Even so, Mekorot was brought in to share its advanced technology with the private contractors to achieve the best outcome.[16]

If water is commonly seen as the domain of government, encouragement of a private sector role is wise industrial policy.

Measure and Monitor

In the mid-1950s, Israel passed a law that no water could flow from a well or into a home, business, or farm without first going through a water meter.[17] Long before Big Data became commonplace—and decades before major cities like London abandoned a flat monthly fee in favor of water meters—Israel began compiling detailed information on water-usage patterns, and analyzing those patterns to detect trends.[18] With this high-level, data-based approach, Israeli water planners have long had the facts necessary to decide whether and when to explore for water, develop resources, and construct facilities—all before the public even knew that such actions were needed.[19]

"If you want to manage your water resources," says Dr. Diego Berger, a hydrologist at Mekorot, "you have to know the consumption patterns of your consumer. Israel knows exactly how much is being used, and for what. With this knowledge, planners can make smart decisions."[20]

Beyond big decisions like expanding exploration, Israeli utilities have been able to employ usage patterns to detect anomalies suggestive of a leak. When the usage pattern seems suspicious, an alert is promptly sent. If the homeowner or landlord was filling a tank or a swimming pool, the file is closed. The water is being used as intended. But when there is no known explanation, a repair crew can get to work immediately. This not only saves consumers on their water bills, but keeps the water lost to leaks as low as possible.[21]

Israel doesn't just keep track of quantity, it casts a similarly wide net in closely following and compiling an array of information about quality. "With all of the data we now have on the water in our system," says Yossi Shmaya, the senior Mekorot executive, "we can anticipate a problem before it actually becomes one. We know what is normal at a given time of year or at a certain temperature."

Shmaya is tight-lipped about how Israel's water security is managed, but says, "Israel isn't the only country that needs to be worried about attacks targeting water. Toxins can get into the water from many sources and not just from terrorists. Everyone needs to have systems in place to identify threats. Everyone has to be able to immediately stop the flow of contaminated water and replace it with a confirmed pure source so that consumers never have to think twice about the safety of their water."[22]

Other countries measure and monitor their water, says Diego Berger. "What makes Israel unique is the comprehensiveness and integration," he says. "The more data you have and the more early-warning systems in place, the easier it is to integrate all parts of your water supply."[23]

Plan Today for Long into the Future

In only the past few decades, all around the world, aquifers that may have taken thousands or millions of years to fill have been depleted by overpumping or tainted with chemicals. The farmers and cities dependent upon these underground reservoirs will soon need to either substantially slow their withdrawals—even with the economic cost that will cause—or find other sources of water.

Due to a lack of long-range planning, these aquifers are now at risk. Israel has been developing rolling master plans since the 1930s and has already been at work for several years on its water plan for 2050. "In Israel," says Michael "Miki" Zaide, the head of strategic planning at the Israel Water Authority, "everyone must coordinate according to the master plan. Around the world, many places have master plans. We aren't unique in that, but these other plans are rarely mandatory. Here, we plan and then strictly implement it."[24]

"Israel engages in excellent planning," says Menachem Priel, the head of the desalination department at Mekorot. "But it goes beyond planning. We also put into our plans how we can make use of the edge of technology. We integrate planning and new approaches and ideas. Because we plan so far out, we can think about technology and infrastructure that might be needed that doesn't yet exist. Looking out for decades gives us time to develop and integrate these ideas."[25]

There will always be the potential for surprises. Long droughts, population booms, or new water-intensive technologies could add to the water deficit, sometimes abruptly. But planning can also account for periodic shocks so that surprises are less painful. A plentiful water supply can be an illusion if there is no planning as to how much could be depleted. Likewise, water can abruptly go offline if quality standards aren't set and constantly monitored to assure compliance.

While individuals think in terms of months and years, water planners need to plot activity mostly in decades. An aquifer or a lake won't dry up in a year or two, but pollution, overpumping, or climate change can spell irreversible doom for water resources over a generation. A disciplined approach to long-range planning is needed for each generation to pass enhanced water resources to its heirs.

Advocates Needed

In most of the world, water gets little media attention or public comment. Generally, unless there is a gushing, broken water pipe that leads the evening news or a crisis such as a prolonged drought, which is often presented as if water shortages came without warning, media outlets show little interest. Likewise, water issues are rarely a topic of popular concern. An informed citizenry—including business and community leaders and an engaged media—needs to be a part of the planning and solutions to water needs. "Energy issues have high awareness in government," says Pat Mulroy, the longtime head of water for Las Vegas and surrounding areas, "because the energy companies have been able to educate politicians as to their needs. Water is managed by utilities or government agencies and there is no one to educate the politicians as to what water problems lay ahead."[26]

The result, Mulroy says, is that water spending and planning get only a fraction of the government and business attention that energy production gets. To remedy this, she says, water needs its own advocates within government.

In Israel, water has long had advocates at a high level of government. In Israel's early years, the chief advocate for water policy was the prime minister himself, David Ben-Gurion. And Israel's third prime minister, Levi Eshkol, was a cofounder and longtime head of Mekorot, Israel's national water utility.

Today, powerful state institutions like the Israel Water Authority, Mekorot, and farmers' groups have access to other decision makers in government as do many of the municipal water corporations found in over fifty Israeli cities and towns. In addition, the public's water interests formally have a high-ranking government official focused on the issue; one member of the prime minister's Cabinet is responsible for water-management issues. While each of these individually can make the case for responsible water policies, all of them taken together create a powerful and well-connected advocacy group that keeps a focus on adequate funding and planning for water. Unlike elsewhere, this institutionally strong and overlapping water elite assures that the country need not wait for a water crisis before addressing water needs. The popular media in Israel covers water stories, and the public is generally well informed about water concerns.

By having vocal and respected water advocates, water infrastructure gets the focus and funding needed, and entrepreneurs get appropriate incentives to develop water-related technologies. These advocates have helped Israel become a world leader in water technology, management, and governance by assuring that water issues were always at the forefront of policy decision-making at the highest levels of government and society.

The Time to Act Is Now

With a global water crisis looming, the Israeli inclination toward taking bold steps may be the most important contribution of its water philosophy to an increasingly water-starved world. Knowing that hazards often lurk over the horizon, getting ahead of a crisis is a core part of Israeli governance. This mind-set also permeates the world of water in Israel. As a result, since at least the 1930s, the country has been getting ahead of water issues before they could become crises.

When there were only local water sources, the Jewish community began planning and then building national networks to transport water from where it was to where it was most needed. When there was not yet a domestic market for using treated sewage in agriculture, the country began building a national infrastructure to support what has grown to be world-leading usage of reclaimed water. While powerful antagonists were urging against building desalination facilities and when it would have been easiest to continue with existing water policy, the country made the decision to build an expensive desalination plant followed in rapid succession by four others. These bold actions are models for what will be needed to address water shortages around the world.

"If there is an important lesson for others from the Israel experience," says Abraham Tenne, a senior official at the Israel Water Authority and an authority on both desalination and wastewater treatment, "it is to not wait until all the answers are in. We figure things out well enough to start and go into each project knowing that it won't be perfect. It doesn't have to be because we know that we can fix it along the way."

The mentality of waiting until everything is perfect, Tenne says, creates long delays. "Worse," he says, "frequently nothing gets started. The need for water is growing and natural water resources are being depleted with potentially terrible environmental consequences. Not taking action is also a kind of taking action. It is opting for the status quo."[27]

With a water crisis at hand, the time to act is now. Israel has shown how.

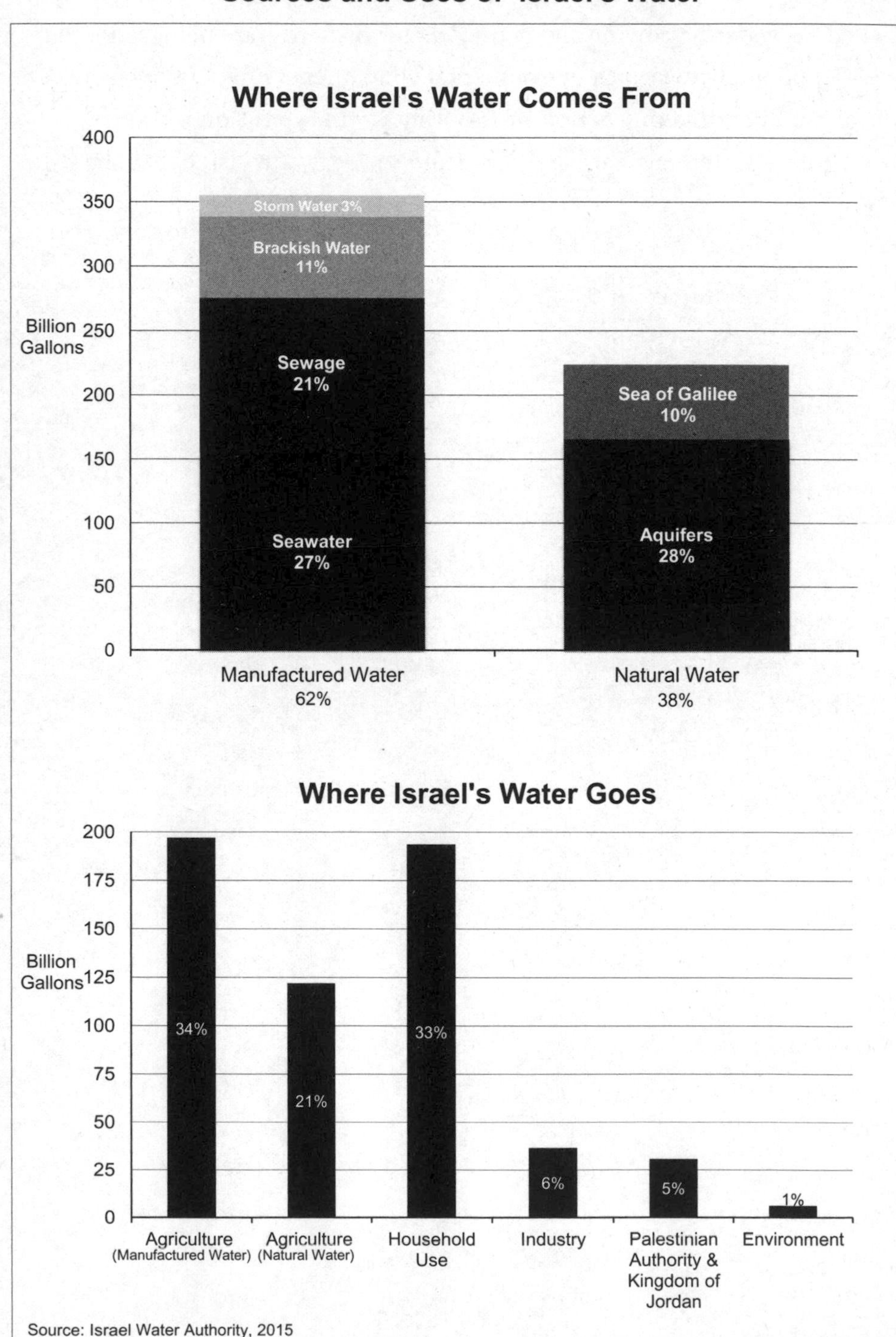

Sources and Uses of Israel's Water
Where Israel's Water Comes From
Billion Gallons
400
350
300
250
200
150
100
50
0
Storm Water 3%
Brackish Water 11%
Sewage 21%
Seawater 27%
Sea of Galilee 10%
Aquifers 28%
Manufactured Water 62%
Natural Water 38%
Where Israel's Water Goes
Billion Gallons
200
175
150
125
100
75
50
25
0
34%
21%
33%
6%
5%
1%
Agriculture (Manufactured Water)
Agriculture (Natural Water)
Household Use
Industry
Palestinian Authority & Kingdom of Jordan
Environment
Source: Israel Water Authority, 2015

Acknowledgments

I am very fortunate to have had so many people assist me in learning about and in telling the story of the coming global water crisis and Israel's role in helping the world to avoid the worst of it.

In all, I interviewed more than 220 people for the book, many of them more than once. I never failed to be amazed at how generous they all were with their time in answering my questions and also in their willingness to introduce me to others. In the main, the interview subjects were Israeli water professionals—officials, regulators, utility executives, professors, businesspeople, entrepreneurs, NGO leaders, engineers, and others. The common thread among them was a great pride in the role that Israel had played in establishing its water independence and in sharing that accumulated expertise with the world. Their excitement and integrity were an inspiration.

Even though everyone I interviewed gave so much to this book, a few stand out. Shimon Tal sat through nine long interviews over more than a year, patiently explaining every facet of Israel's water world. He became a role model for his kindness, patience, and humility. Dr. Moshe Gablinger not only told me remarkable stories from his world travels with TAHAL, but also served as my tutor in geology and other areas of science relating to water. Professor Uri Shani helped me to understand the complexity of both soil science and Israeli-Palestinian-Jordanian relations, among much else. Naty Barak had been involved with every part of the rise of drip irrigation technology, and exuberantly shared all he knew; in the process, our families became friends. Westher Hess, the daughter of Inez and Walter Clay Lowdermilk, helped round out the life stories of this remarkable couple. Yitzhak Blass gave a fuller picture of his father, Simcha Blass, modern Israel's first water genius.

Many helped in other ways. Ido Aharoni, Israel's Consul General in New York, and Nili Shalev, Israel's Economic Minister to North America, each made several introductions to key people in the Israel water world. David Goodtree invited me to join a water seminar in Israel where I met senior water people and also got to visit major water infrastructure. Asaf Shariv not only opened many doors in Israel, but also arranged for the president of Israel, Shimon Peres, to sit for an interview lasting most of a long morning. Oded Distel seems to know everyone in the water business in Israel, and introduced me to many of them.

Russell Robinson and Zevi Kahanov of JNF-USA were an endless source of insight into water issues in Israel. Zevi passed away in the middle of this project and is sorely missed. Doron Krakow from Ben-Gurion University connected me with professors in a variety of disciplines.

Osnat Maron, the archivist at Mekorot, sent me file cabinets full of reports to read, some going back to the early 1950s, and also dozens of photographs to bring those reports to life. Her colleague Udi Zuckerman served as door opener and translator for me with Mekorot's extraordinary staff of professionals as well as being an interview subject himself. Likewise, Olga Slepner helped me to navigate the Israel Water Authority, and with good humor endured my many requests for information and clarifications. Dillon Hosier with the Consul General of Israel's office in Los Angeles shared the background of the Israel-California relationship and made introductions to people in both governments. Carolyn Starman Hessel offered sage advice at crucial milestones. Dan Doctoroff, Phil Lerner, Yana Lukeman, Mike Pevzner, Michael Sonnenfeldt, and Patricia Udell each helped the project in a significant way. My daughter, Talia Siegel, provided design guidance.

So many of those I interviewed inspired me that this was an important story to tell. These include Shmuel Aberbach, Rotem Arad, Diego Berger, Ilan Cohen, Shoshan Haran, Arie Issar, Eugene Kandel, Booky Oren, Huageng Pan, Chemi Peres, Sandra Shapira, Yossi Shmaya, Tami Shor, Abraham Tenne, Ronen Wolfman, and Sivan Ya'ari, among many others.

I was also fortunate to have several people volunteer to read and comment on the book. Sam Adelsberg, Laureine Greenbaum, Oliver Herzfeld, Dan Polisar, Peter Rup, and my son, Sam Siegel, provided invaluable comments causing me to rethink assumptions, rewrite sections and whole chapters, and reorganize the flow of the book. For each of these six people, I am awed by the time and focus they put into the manuscript. The book is far better for their ideas and input.

In addition to those who read the entire book, subject matter experts critiqued specific chapters. Professor Tuvia Friling read the chapter on the National Water Carrier, Professor Ronnie Friedman and Naty Barak on drip irrigation, civil engineers Ely Greenberg and Teddy Fischer on the treatment and reuse of sewage, Professor Rafi Semiat on desalination, and Dr. David Pargament and Dr. Clive Lipchin on rivers. Each added nuance and clarity.

Donna Herzog, a Ph.D. candidate in Israel Studies, read nearly every chapter in which historical material was included, and made profound and insightful comments. She pointed me to archival collections and taught me how to make use of them. She also translated many important documents from Hebrew to English. Some wise university will add her to its faculty.

Given the political environment surrounding the Israeli-Palestinian issue, I was eager to learn about mutual Israeli and Palestinian claims, grievances, and, most of all, opportunities for cooperation. Former Palestinian Prime Minister Salam Fayyad generously let me interview him three times. Almotaz Abadi of the Palestinian Water Authority not only answered many questions and referred me to reports, but also set up interviews with Palestinian professors, other water officials, and heads of NGOs. I never met Professor Yousef Abu Mayla—who lives in Gaza—but we have spoken and emailed many times. He kindly read successive drafts of the Palestinian chapter and made, in total, dozens of comments that greatly enriched my understanding. On the Israeli side, among others, Professor Eilon Adar, Gidon Bromberg, Ambassador Nadav Cohen, and Dr. Clive Lipchin answered countless questions, each over several sessions. Many thanks, as well, to Dave Harden for both his service to those in need and the insights he shared.

I can't imagine a first-time author getting luckier than I have been. My agent, Mel Berger of William Morris Endeavor, was excited by the project just from the description and swiftly sold the book. Marcia Markland, my editor (and the person who acquired the book), has been a great enthusiast of the project and, despite the mayhem it must have caused to her scheduling, trusted my judgment in taking back completed sections and completely rewriting them. She has also become a friend. There aren't enough good things to say about the people at Thomas Dunne Books and St. Martin's Press. Thank you to Jeff Capshew, Laura Clark, Tom Dunne, Tracey Guest, Alastair Hayes, Kathryn Hough, Quressa Robinson, and Pete Wolverton for such a wonderful experience.

When I hired Jamie Black as my research assistant, I could never have imagined that anyone would have felt a sense of ownership in the book equal to my own. A young man of exceptional intellectual gifts, he is also a master of organization and technology, using these skills daily on this project. Tenacious and always driven for perfection, Jamie would push himself and me to continually rethink some element of the manuscript or research. I look forward to working with him for years to come.

Finally, I want to acknowledge my family, all of whom, but especially my wife, have heard many, many water stories over the past few years. More than any other element, our children, Alana, Sam, and Talia, bring joy into our lives. My wife and their mother, Rachel Ringler, is my friend, partner, and inspiration. Every day, I am awed by her wisdom, kindness, and beauty. In acknowledgement of that, I dedicate this book to her.

INTERVIEW LIST

(Titles shown are accurate as of the time of the interview)

NOTE: For different reasons, some of those interviewed wished to be anonymous. Since that "Not for Attribution" requirement aided in their providing candor and context, their names do not appear here.

Rashad Al-Sa'ed—Professor, Sanitary and Environmental Engineering, Birzeit University

Almotaz Abadi—Advisor, Head of Aid Coordination Unit, Palestinian Water Authority

Alfred Abed Rabo—Professor, Chemistry and Geohydrology, Bethlehem University

Shmuel Aberbach—Retired, Deputy Director of Hydrology, TAHAL

Itzhak Abt—Former Director, Agriculture, Water & Environment department, Peres Center for Peace; former Director, CINADCO, Ministry of Agriculture

Yousef Abu Mayla—Professor, Water and Environmental Studies, Al Azhar University, Gaza

Eilon Adar—Director, Zuckerberg Institute for Water Research; Professor, Hydrogeology and Arid Zones Research, Ben-Gurion University

Refael Aharon—CEO, Applied Cleantech

Avi Aharoni—Director, Wastewater and Reuse Department, Mekorot

Ido Aharoni—Consul General of Israel (New York)

Zvi Amit—Manager, Seed Breeding, Hazera Genetics Ltd.

Rotem Arad—VP, Food and Beverage for USA/Europe/Japan, Atlantium

Ruti Arad—Daughter of Aaron Wiener

Uzi Arad—Son-in-law of Aaron Wiener; former Israel National Security Advisor
Shaul Arlosoroff—Former Director, Mekorot; former Deputy Water Commissioner
Natan Aridan—Editor, Israel Studies; Ben-Gurion University
Danny Ariel—Sugarcane and Smallholder Product Manager, Netafim
David Arison—Director, Global Business Relations, Miya Arison Group
Shaul Ashkenazy—Chairman of the Board, Plasson; former General Manager, Plasson
Assaf Atar—Co-archivist, Kibbutz Hatzerim
Shaddad Attili—Palestinian Authority Minister of Water; Chairman, Palestinian Water Authority
Ornit Avidar—Founder and Managing Director, Waterways Solutions
Ram Aviram—Professor of Water Policy, Tel Hai College; former Israel Ambassador, Water Diplomacy
Dror Avisar—Professor, Hydrochemistry and Director, Hydrochemistry Laboratory, Tel Aviv University
Yehuda Avner—Author, "The Prime Ministers"; former Israel Ambassador
Bonnie Azoulay—Aquatic Biologist, Mekorot
Miriam Balaban—Founder and Editor, Desalination and Water Treatment Journal
Moshe Bar—Global Director, External Collaboration for Seed Breeding, Syngenta
Naty Barak—Chief Sustainability Officer, Netafim
Nir Barlev—Managing Director, Ra'anana Water Corporation
Assaf Barnea—CEO, Kinrot Ventures
Jehad Bashir—Head, Joint Technical Committee, Palestinian Water Authority
Sarit Bason—Desalination Engineer, Mekorot
Assaf Bassi—R&D Manager, Galcon
Erika Ben-Basat—Research Coordinator, R&D Department, Amiad

Ronen Benjamin—COO, Owini, a division of Mitrelli; Angola project director
Diego Berger—Kinneret Watershed Hydrologist, Mekorot
Nathan Berkman—Co-founder, ADAN Technological and Economic Services Ltd.; former Managing Director, IDE Technologies
Yitzhak Blass—Son of Simcha Blass
Oren Blonder—Vice President, Sales & Marketing, MemTECH
Saar Bracha—CEO, TAHAL
Shmuel Brenner—Former Deputy Director, Israel Ministry of Environment; Professor, Arava Institute of Environmental Studies
Gidon Bromberg—Israel Co-Director, EcoPeace Middle East
Ilan Cohen—Former Director, Budget Department, Ministry of Finance; business strategist
Nadav Cohen—Ambassador, Water Diplomacy, Israel Ministry of Foreign Affairs
Gabby Czertok—Former CEO, HydroSpin
Arik Dayan—President and CEO, Amiad
Oded Distel—Director, Israel NewTech, Israel Ministry of Economy
Haim Divon—Israel Ambassador to The Netherlands; former director, The Center for International Cooperation (MASHAV), Israel Ministry of Foreign Affairs
Ze'ev Efrat—CEO, Aquarius Spectrum
Sara Elhanany—Director, Water Quality Division, Israel Water Authority
Richard Engel—Director, Environment and Natural Resources Program, National Intelligence Council; Major General, U.S. Air Force (ret.)
Miriam Eshkol—Widow of former Israeli Prime Minister Levi Eshkol
Yaqub Eyad—Director, Solid Waste and Wastewater for Salfit, West Bank; former official, Palestinian Water Authority
Salam Fayyad—Former Prime Minister, Palestinian Authority
Gilad Fernandes—Senior Deputy Director General (Economics), Israel Water Authority
Ze'ev Fisher—Founder and CEO, Mapal Green Energy Ltd.

Oded Fixler—Senior Deputy Director General (Engineering), Israel Water Authority

Itai Freeman—CEO, Pareto Strategies; former Manager, Besor River Restoration; Development Manager, Beersheba River Park

Elad Frenkel—CEO, Aqwise

Ronnie Friedman—Professor of Immunology, Hebrew University of Jerusalem (Faculty of Agriculture)

Tuvia Friling—Professor, Israel History, Ben-Gurion University

Noah Galil—Professor of Civil and Environmental Engineering, Technion—Israel Institute of Technology

Moshe Ga'on—CEO, Gaon Holdings

Moshe Gablinger—Retired, Senior Water Resources Engineer, TAHAL

Assaf Gavron—Novelist, author of "Hydromania"

Shlomo Getz—Director, Institute for the Research of the Kibbutz, University of Haifa

Shabtai Glass—Manager, Water Association of the Upper Galilee

Guilford Glazer—Philanthropist (Ben-Gurion's guide on his visit to the Tennessee Valley Authority)

Nati Glick—Director, Rain-Cloud Seeding, EMS—Mekorot Projects

Noam Goldstein—VP, Potash Operations, Dead Sea Works

Noam Gonen—Partner, Trigger-Foresight, a division of Deloitte Consulting (author of the "Agamim" study)

Don Gopen—Shafdan Project Manager, CDM

Haim Gvirtzman—Professor, Hydrology, Institute of Earth Sciences at Hebrew University of Jerusalem; member, Israel Water Authority Council

Kobi Haber—Director, Finance and Economic Division, Bank Leumi; former Director, Budget Department, Ministry of Finance

Dvora HaCohen—Professor, Modern Jewish History, Bar-Ilan University

Suleiman Halasah—Acting Associate Director, Center for Transboundary Water Management, The Arava Institute for Environmental Studies

Shoshan Haran—Founder and General Manager, Fair Planet; former Director of Trait Exploration, Hazera Genetics Ltd.

Dave Harden—Mission Director, USAID West Bank and Gaza

Leila Hashweh—Graduate Student, Ben-Gurion University

Zvi Herman—Former Director, Center for International Agriculture Development Cooperation (CINADCO), Ministry of Agriculture

Donna Herzog—Ph.D. candidate, Israel Studies, New York University

Westher Hess—Daughter of Inez Marks Lowdermilk and Walter Clay Lowdermilk

Daniel Hoffman—Founding Co-Partner, ADAN Technological and Economic Services Ltd.

Dillon Hosier—Political Advisor to the Consul General of Israel (Los Angeles)

Avshalom Hurvitz—Biologist and R&D Manager, GalilAlgae (fish ponds)

Nelly Icekson-Tal—Director, Shafdan Wastewater Treatment Facility, Mekorot

Rafi Ifergan—Chief Technology Officer, Mekorot

Eytan Israeli—Consultant, Israel Water Works Association

Ran Israeli—CEO, Bermad

Arie Issar—Professor Emeritus, Environmental Hydrology and Microbiology, Ben-Gurion University

Paula Kabalo—Director, Ben-Gurion Research Institute for the Study of Israel and Zionism, Ben-Gurion University

Zevi Kahanov—Director, JNF Parsons Water Fund

Adam Kanarek—Former General Manager, Shafdan Wastewater Treatment Facility, Mekorot

Eugene Kandel—Head, National Economic Council, Prime Minister's Office, Israel

Itzik Kantor—Former General Manager, Plasson

Yoram Kapulnik—Director, Agricultural Research Organization, Volcani Center

Fadil Kawash—Former Commissioner, Palestinian Water Authority; former Chief Palestinian negotiator for water issues

Ora Kedem—Professor Emerita, Weizmann Institute; co-founder, Ben-Gurion University

Ruth Keren—Chief Archivist, Kibbutz Hatzerim

Kris Kiefer—General Counsel, Office of U.S. Senator Jeff Flake (Arizona)

Marcus King—Director of Research, Elliot School of International Affairs, George Washington University

Yoav Kislev—Professor, Agricultural Economics, Hebrew University of Jerusalem

Karni Krieger—Office of the CEO, Mekorot

Naomi Lauter—Friend of the Lowdermilk family

Eytan Levy—CEO, Emefcy

Clive Lipchin—Director, Center for Transboundary Water Management, Arava Institute of Environmental Studies

Miriam "Mickey" Loeb—Widow of Sidney Loeb

Fredi Lokiec—Executive Vice President, Special Projects, IDE Technologies

Uri Lubrani—Former Israel Ambassador to Iran

Joe MacIlvaine—President, Paramount Farming Company

Israel Mantel—Retired, Senior Executive, Mekorot

Doron Markel—Director, Monitoring and Management, Lake Kinneret and its Watershed, Israel Water Authority

Sondra Markowitz—Friend of Inez Marks Lowdermilk

Osnat Maron—Chief Archivist, Mekorot

Bashar Masri—Founder, Massar International

Raphael "Rafi" Mehoudar—Inventor

Erez Meltzer—Former CEO, Netafim

Clemens Messerschmidt—Hydrogeologist

Medy Michail—Former Consulting Engineer, TAHAL

Hillel Milo—Co-CEO, AquaAgro Fund

Aya Mironi—Executive, Google

Pat Mulroy—General Manager, Southern Nevada Water Authority (SNWA)

Baruch Nagar—Director, Water Administration for the West Bank and Gaza, Israel Water Authority

Efi Naim—Hula Valley and Golan District Ranger (Northern Region), KKL-JNF

Shahar Nuriel—Technologist, R&D Department, Amiad

Booky Oren—CEO, Booky Oren Global Water Technologies Ltd.; former Chairman, Mekorot; former Director, Business Development, Netafim

Gili Ovadia—Director, Israel Economic Mission, West Coast, USA

Huageng Pan—Founder and President, Dowell Technological & Environmental Engineering Co.

Tom Pankratz—Editor, Water Desalination Report

David Pargament—Director General, Yarkon River Authority

Amir Peleg—Founder and CEO, TaKaDu

Chemi Peres—CEO, Pitango

Shimon Peres—President, State of Israel

Mark Peters—Senior Technical and Policy Advisor, USAID West Bank and Gaza

Menahem Priel—Director, Desalination and Special Projects Division, Mekorot

Kish Rajan—Director, California Governor's Office of Business and Economic Development

Mira Rashty—CEO, Tel Afek—The National Water Center; owner, Sippur Pashut bookstore

Sari Razili—COO, Galcon

Ariel Rejwan—VP, Business Development, Mekorot

Dan Reisner—Partner, Herzog Fox & Neeman; Chief Water Negotiator, Israel-Palestinian Authority Negotiations; Draftsman, Oslo II Accords Water Articles, and Israel-Jordan Water Agreement

Guy Reshef—Water Quality and Monitoring Supervisor, Israel Water Authority

Edward Rifman—Major, Israel Air Force Construction Unit, Ramon Air Force base

Russell Robinson—CEO, JNF (USA)

Meir Rom—Statistician, Lake Kinneret Watershed Unit, Mekorot

Rami Ronen—CEO, Strauss Water

Taniv Rophe—Deputy Director of Planning, Israel Ministry of Agriculture and Rural Planning

Ruhakana Rugunda—Prime Minister, Uganda

Mary Rose Ryan—Friend of Inez Marks Lowdermilk and Walter Clay Lowdermilk

Ofer Sachs—CEO, The Israel Export and International Cooperation Institute

Ariel Sagi—CEO, ARI

David "Dudu" Sapir—Water Supply Engineer, Mekorot

Uri Schor—Spokesperson, Israel Water Authority

Yehoshua Schwarz—Consultant, TAHAL; former senior executive, TAHAL

Raphael Semiat—Professor, Chemical Engineering, Technion—Israel Institute of Technology

Nechemya Shahaf—General Director, Besor River Authority

Ami Shaham—General Manager, Arava Drainage Authority

Gabi Shaham—Water consultant; former engineer, TAHAL

Amos Shalev—Chairman, Bermad

Nili Shalev—Israel Economic Minister to North America

Yosi Shalhevet—Former Director, Agricultural Research Organization, Volcani Center; former representative of the Israel Academy of Sciences in China

Yossi Shmaya—District Manager, Central Filtration Facility, Mekorot

Uri Shamir—Professor Emeritus, Technion—Israel Institute of Technology; Founding Director, Grand Water Research Institute

Uri Shani—Former Chairman, Israel Water Authority; Professor, Hebrew University; Businessman; Inventor

Asaf Shariv—Co-Founder and Partner, Amelia Ventures; former CEO, Peres Center for Peace; former Consul General of Israel (New York)

Sandra Shapira—Director, Corporate Communications, Amiad

Rachel Shaul—Director, Corporate Marketing, Netafim

Amnon Shefi—CEO, Hi-Teach

Avraham Baiga Shochat—Former Israel Minister of Finance

Barbara Shivek—Co-archivist, Kibbutz Hatzerim

Tami Shor—Deputy Director General (Regulation), Israel Water Authority

David Siegel—Consul General of Israel (Los Angeles)

Davide Signa—Food Security Expert, UN Food and Agriculture Organization (FAO)

Olga Slepner—Director, Director-General's Bureau and Director of the Foreign Relations Unit, Israel Water Authority

Yossi Smoler—Director, Technological Incubator Program, Office of the Chief Scientist, Israel Ministry of Economy

Ephraim Sneh—Former Deputy Minister of Defense; former Head, West Bank Civil Administration

Arnon Soffer—Professor Emeritus and Founder, Department of Geography, University of Haifa

Jeffrey Sosland—Professor, American University; author, "Cooperating Rivals: The Riparian Politics of the Jordan River Basin"

Efi Stenzler—CEO and Chairman, KKL

Andrew Stone—Member of the British House of Lords; former Managing Director, Marks and Spencer

Kelvin Stroud—Legislative Aide, Senate Water Caucus in the office of Senator Mark Pryor (Arkansas)

Charles Swartz—Program Manager, US-Israel Science and Technology Foundation; former program manager, MERC

Alon Tal—Professor, Environmental Policy, Ben-Gurion University; founder, Arava Institute; Chairman, Israel Green Party

Danny Tal—Commissioner for Trade Levies, Israel Ministry of Economy; former Israel Economic Counselor, Guangdong Province, China (2008–2013)

Shimon Tal—Former Israel Water Commissioner; former Chief Technology Officer, Mekorot

Abdelrahman Tamimi—Director, Palestinian Hydrology Group for Water and Environmental Resources Development

Micha Taub—Operations Director, Soreq Desalination Facility, IDE Technologies

Abraham Tenne—Director, Desalination Division, Israel Water Authority; chairman, Desalinated Water Administration

Nirit Ulitzer—Co-Founder and Chief Technology Officer, CheckLight

David Waxman—Chairman, NiroSoft; former CEO, IDE Technologies

Alex Weisberg—Former Vice-Director, Water Supply, TAHAL

Yirmi Weisberg—Former Vice President, TAHAL

Uri Werber—Founder and former General Manager, Netafim

Ronen Wolfman—CFO and Board Member, Hutchison Water Group; former CEO, Mekorot; former Deputy Director of Budget for Infrastructure, Israel Ministry of Finance

Sivan Ya'ari—Founder and CEO, Innovation: Africa

Glenn Yago—Founder, Financial Innovation Lab, Milken Institute

Meir Yaacoby—Chief Engineer, Innovation: Africa

Yossi Yaacoby—Director, WaTech, Mekorot

Humberto Yaakov—Manager, Yad Hana Wastewater Treatment Facility

(Colonel) Grisha Yakubovich—Director of Civil Coordination Department, Coordination of Government Activities in the Territories (COGAT), Israel Ministry of Defense

Adi Yefet—Director, Water Department, NewTech, Israel Ministry of Economy

Alon Yegens—Executive Vice President, TAHAL International; CEO, TAHAL Assets

Barak Yekutiely—Founder and CEO, Aquate

Joshua Yeres—Senior Advisor for Business Development, Hagihon-Jerusalem Water and Sewage Utility

Zohar Yinon—CEO, Hagihon-Jerusalem Water and Sewage Utility

Moshe Yizraeli—Consultant, regional water issues, Israel Water Authority

Ori Yogev—Chairman, WhiteWater; former Director, Budget Department, Ministry of Finance

Michael "Miki" Zaide—Manager, Strategic Planning, Israel Water Authority

Dan Zaslavsky—Professor Emeritus, Agricultural Engineering, Technion—Israel Institute of Technology; former Chairman, Israel Water Commission

Jim Zehringer—Director, State of Ohio Department of Natural Resources

Aviram Zuck—Upper Galilee and Golan Heights—Forest Supervisor, KKL-JNF

Udi Zukerman—Director, Global Business and Entrepreneurship, Mekorot

Notes

INTRODUCTION: A GLOBAL WATER CRISIS LOOMS

1. "Who We Are," National Intelligence Council, accessed on November 24, 2014: www.dni.gov/index.php/about/organization/national-intelligence-council-who-we-are.
2. Office of the Director of National Intelligence, National Intelligence Council, *Global Water Security*, Intelligence Community Assessment (Washington, DC: National Intelligence Council, February 2, 2012).
3. Office of the Director of National Intelligence, National Intelligence Council, *Global Trends 2025: A Transformed World* (Washington, DC: National Intelligence Council, November 2008), 51.
4. Office of the Director of National Intelligence, *Global Water Security*, *op. cit.*, iv.
5. Ibid., iii.
6. Luciana Magalhaes, Reed Johnson, and Paul Kiernan, "Blackouts Roll through Large Swath of Brazil," *The Wall Street Journal*, January 19, 2015.
7. Office of the Director of National Intelligence, *Global Water Security*, *op. cit.*, iii.
8. For a broader discussion of global water security and the intelligence community assessment of it, see: Marcus DuBois King, *Water, U.S. Foreign Policy and American Leadership* (Washington, DC: Elliott School of International Affairs, George Washington University, October 2013).
9. U.S. Department of Interior, U.S. Geological Survey, *Ground-Water Availability in the United States*, eds. Thomas E. Reilly, Kevin F. Dennehy, William M. Alley, and William L. Cunning (Reston, VA: U.S. Geological Survey), 44.

10. For an example where the High Plains Aquifer has run dry, see NBC News profile of a farm in Amarillo, Texas, that is no longer able to pump water from the once plentiful aquifer below. [Brian Brown, "The Last Drop: America's Breadbasket Faces Dire Water Crisis," *NBC News*, July 6, 2014.]
11. Erla Zwingle, "Ogallala Aquifer: Well Spring of the High Plains," *National Geographic* 183, no. 3, 83.
12. U.S. Department of Interior, U.S. Geological Survey, *Groundwater Depletion in the United States (1900–2008)*, Scientific Investigations Report 2013-5079, ed. Leonard F. Konikow (Reston, VA: U.S. Geological Survey, 2013), 22.
13. Michael Wines, "States in Parched Southwest Take Steps to Bolster Lake Mead," *The New York Times*, December 17, 2014.
14. In February 2014, Williams, Arizona, enacted water restrictions so severe, they included a ban on potable or raw water "for any purpose other than public health or emergency purposes" and stopped all new building permits. [City of Williams, *Level 4 Water Restrictions: URGENT NOTICE* (Williams, AZ: City of Williams, 2014).]
15. Lizette Alvarez, "Florida Lawmakers Proposing a Salve for Ailing Springs," *The New York Times*, April 14, 2014.
16. United Nations, Department of Economic and Social Affairs, *World Population Prospects: The 2012 Revision, Key Findings and Advance Tables* (New York: United Nations, Department of Economic and Social Affairs, 2013), 1.
17. Over the next four decades, world population is forecast to increase by two billion people to exceed nine billion people by 2050. Recent estimates by the United Nations Food and Agriculture Organization (FAO) indicate that to meet the projected demand, global agricultural production will have to increase by sixty percent from its 2005–2007 levels. [Food and Agriculture Organization of the United Nations, *Statistical Yearbook 2013, World Food and Agriculture* (Rome: Food and Agriculture Organization of the United Nations, 2013), 123.]
18. Homi Kharas and Geoffrey Gertz, *The New Global Middle Class: A Cross-Over from West to East* (Washington, DC: Wolfensohn Center for Development at Brookings, 2010), 5.
19. Office of the Director of National Intelligence, *Global Water Security, op. cit.*, i.
20. According to a 2014 United Nations report, "[t]ypical volumes of water needed end-to-end (from extraction through refining) for petroleum based fuels are [2 to 4 gallons of] water per [gallon of] fuel. For natural gas, the volumes of water are approximately [5 to 13 gallons of] water per barrel equivalent of oil. . . . For hydraulic fracturing, typical water injection volumes are [2 to 8] million [gallons] per well." [United Nations World Water Assessment Program, *The United Nations World Water Development Report 2014: Water and Energy* (Paris: United Nations Education, Scientific and Cultural Organization, 2014), 30.]

21. The Great Lakes are an example where higher surface water temperature has caused water loss through greater evaporation. [Andrew D. Gronewold and Craig A. Stow, "Water Loss from the Great Lakes," *Science* 343 (March 7, 2014): 1084–1085.]
22. Doron Markel, author's interview, Sapir Pumping Station (Israel), April 29, 2013.
23. As an example, over a thirty-year period, as many as five hundred thousand soldiers and civilians in Camp Lejeune, North Carolina may have consumed water contaminated with carcinogenic agents that came from an off-base dry cleaning company and other sources. [U.S. Department of Health and Human Services, The President's Cancer Panel, *Reducing Environmental Cancer Risk: What We Can Do Now: 2008–2009 Annual Report President's Cancer Panel*, ed. Suzanne H. Reuben (Bethesda, MD: President's Cancer Panel, 2010), 78.]
24. Smart Water Networks Forum Research, *Stated NRW (Non-Revenue Water) Rates in Urban Networks* (Portsmouth, U.K.: Smart Water Networks Forum, August 2010), 3.
25. Amman's average water loss between 1986 and 2004 was 53 percent.* Adana, Turkey's fifth biggest city, has a water loss of 69 percent.**

    * Nadhir Al-Ansari, N. Alibrahiem, M. Alsaman, and Sven Knutsson, "Water Supply Network Losses in Jordan," *Journal of Water Resource and Protection* 6, no. 2 (February 2014): 87.

    ** Smart Water Networks Forum Research, *op. cit.*
26. David Dunlap, "Far, Far Below Ground, Directing Water to New York City Taps," *The New York Times*, November 19, 2014.
27. Israel's population at the end of 2014 was 8.3 million.* On May 15, 1948, its population was 806,000.**

    * Central Bureau of Statistics, *Israel's Population on the Eve of Independence Day* (Jerusalem: Central Bureau of Statistics, May 1, 2014).

    ** Central Bureau of Statistics, *Israel in Statistics: 1948–2007* (Jerusalem: Central Bureau of Statistics, May 2009), 2.
28. According to the World Bank, Israel's GDP (current US$) has more than doubled since 2005.
29. Benjamin Netanyahu, *Speech after Signing the California-Israel Cooperation Agreement* (Mountain View, CA, March 5, 2014).
30. Israel was ranked by the World Resources Institute in 2014 as the twenty-first most water-stressed country in the world. It falls into the highest bracket—"Extremely high stress." [Paul Reig, Andrew Maddocks, and Francis Gassert, "World's 36 Most Water-Stressed Countries," *World Resources Institute*, December 12, 2013.]
31. Israel supplies the Palestinian Authority with fifteen billion gallons of water in the West Bank* and 2.6 billion gallons in Gaza.** Israel also currently supplies the Kingdom

of Jordan with 14 billion gallons annually. Furthermore, Israel and the Kingdom of Jordan also signed an agreement in February 2015, where Jordan agreed for Israel to buy 9 billion gallons of desalinated water per year from a newly planned desalination plant in Aqaba, in exchange for Israel doubling its supply to Jordan to 28 billion gallons.***

* Mekorot, *Mekorot's Association with the Palestinians Regarding Water Supplies* (Tel Aviv: Mekorot, 2014), 6. Also see Palestinian Water Authority, *Annual Status Report on Water Resources, Water Supply, and Wastewater in the Occupied State of Palestine—2011* (Ramallah: Palestinian Water Authority, 2012), 28.

** Yaakov Lappin, "Israel to Double Amount of Water Entering Gaza," *The Jerusalem Post*, March 4, 2015.

*** Sharon Udasin, "Israeli, Jordanian Officials Signing Historic Agreement on Water Trade," *The Jerusalem Post*, February 26, 2015.

CHAPTER 1: A WATER-RESPECTING CULTURE

1. Aya Mironi, author's interview, New York, February 3, 2014.
2. Uri Schor, author's interview, Tel Aviv, April 25, 2013.
3. For a partial list of Jewish religious sources in connection with rain and water, see: Wolf Gafni, Pinhas Michaeli, Ahouva Bar-Lev, Yerahmiel Barylka, and Edward Levin, *Beside Streams of Waters: Rain and Water in the Prayers and Ceremonies of the Holiday* (Jerusalem: Jewish National Fund, Keren Kayemeth LeYisrael, Religious Organizations Department, 1990).
4. Num. 20:1–13 and Ex. 17:16.
5. Deut. 11:14 and 28:12.
6. Deut. 11:17.
7. James Strong, *Strong's Exhaustive Concordance of the Bible* (Peabody, MA: Hendrickson Publishers, 2009).
8. For more on Theodor Herzl, see: Amos Elon, *Herzl* (New York: Holt, Rinehart and Winston, 1975) and Shlomo Avineri, *Herzl: Theodor Herzl and the Foundation of the Jewish State*, trans. Haim Watzman (London: Weidenfeld & Nicolson, 2013).
9. Yehuda Avner, *The Prime Ministers: An Intimate Narrative of Israeli Leadership* (New Milford, CT: The Toby Press, 2010), 105.
10. Theodor Herzl, *The Complete Diaries of Theodor Herzl*, ed. Raphael Patai, trans. Harry Zohn (London: Herzl Press, 1960), 755.
11. Theodor Herzl, *Old New Land (Altneuland)* (Minneapolis, MN: Filiquarian Publishing LLC, 2007), 51.

12. Ibid., 264.
13. Ibid., 268.
14. Isa. 12:3.
15. Other choreographed songs about water include: *"Yasem Midbar Le'Agam Mayim"* [A Desert Shall Be Turned to a Lake], 1944, and *"Etz HaRimon"* [The Pomegranate Tree], 1948.
16. Abraham B. Yehoshua, *Early in the Summer of 1970* (New York: Schocken, 1971).
17. Amos Oz, *My Michael* (New York: Knopf, 1972).
18. Assaf Gavron, author's interview, telephone, July 16, 2014.
19. State of Israel, *Water Drilling Control Law*, 5715-1955, Section 4.
20. State of Israel, *Water Measurement Law*, 5715-1955, Section 2(a).
21. Ibid., Section 3(a).
22. State of Israel, *Drainage and Flood Control Law*, 5718-1957, Section 1.
23. Ibid., Section 4(a).
24. Ibid., Section 5.
25. State of Israel Water Commission, *The Water Laws of Israel*, ed. M. Virshubski (Tel Aviv: State of Israel Water Commission, March 1964), i.
26. State of Israel, *Water Law*, 5719-1959, Section 1.
27. Ibid., Section 4.
28. Ibid., Section 3.
29. Ibid., Section 9(1).
30. Shimon Tal, author's interview, telephone, March 11, 2013.
31. French Republic, *Civil Code*, Article 642.
32. Ibid., Article 641.
33. Arnon Soffer, author's interview, Haifa, May 2, 2013. Since former Israeli Prime Minister Ehud Barak had famously used the phrase "a villa in a jungle," but in another context, Professor Soffer was likely borrowing Barak's phrase and adapting it to his comments about water.

CHAPTER 2: THE NATIONAL WATER CARRIER

1. Michael Makovsky, *Churchill's Promised Land: Zionism and Statecraft* (New Haven, CT: Yale University Press, 2008), 183–184.
2. In addition to fears of a Muslim uprising or a "fifth column" in India, Palestine, or other areas under British control, the British were also mindful that a large amount of the oil they would need for a coming war effort was in Iraq and other places with majority Muslim populations. [Tuvia Friling, *Arrows in the Dark: David Ben-Gurion, the Yishuv*

Leadership, and Rescue Attempts during the Holocaust, trans. Ora Cummings (Madison, WI: University of Wisconsin Press, 2005), 2.]

3. For a copy of the British White Paper of 1939, see: Charles D. Smith, *Palestine and the Arab-Israeli Conflict*, 6th ed. (Boston: Bedford/St. Martin's, 2007), 165–169.
4. For a comprehensive study on the question of Palestine's "economic absorptive capacity," see: Shalom Reichman, Yossi Katz, and Yair Paz, "The Absorptive Capacity of Palestine, 1882–1948," *Middle Eastern Studies* 33, no. 2 (1997): 338–361.
5. Miriam Eshkol, author's interview, telephone, April 29, 2013.
6. A few examples of the pre-State entities include the Histadrut Labor Federation, which was founded in 1920. Keren Hayesod, an organization created to raise funds to build state institutions such as hospitals and schools, was also begun in 1920. Solel Boneh was founded in 1921 to pave roads and construct guard towers, and Bank Hapoalim [Workers' Bank] was created that same year. Hebrew University was established in 1925. The Jewish Agency for Palestine was created in 1929 to facilitate immigration and to help absorb the arriving immigrants.
7. The naming of Mekorot had something of a comic element to it. As was common in *Yishuv* organizations at the time, the founders wanted to make use of a biblical name tying the Zionist enterprise to its ancient roots. A company board member found a verse from Psalms that says "the voice [Mekolot] of God is greater than water" (Psalms 93:4), but, mistranscribed Mekolot as Mekorot since Mekorot means "sources," a good name for a water exploration company. When a more biblically literate board member pointed out the misreading of the text at a subsequent company meeting, the first board member stood by his initial proposal and won the argument. [Assaf Seltzer, *Meḳorot: Sipurah Shel Hevrat Ha-Mayim Ha-Le'umit—75 Ha-Shanim Ha-Rishonot* [Mekorot: The Story of the Israel National Water Company—The First 75 Years] (Jerusalem: Yad Yitzhak Ben-Zvi, 2011), 35.]
8. Ibid., 30–32.
9. Haim Gvirtzman, *Mash'abe ha-mayim be-Yiśrael: peraḳim be-hidrologyah uve-mada'e ha-sevivah* [Israel Water Resources: Chapters in Hydrology and Environmental Sciences] (Jerusalem: Yad Ben-Zvi Press, 2002), 190.
10. Aharon Kellerman, *Society and Settlement: Jewish Land of Israel in the Twentieth Century* (Albany, NY: State University of New York Press, 1993), 245–247.
11. Simcha Blass, *Mei Meriva u-Ma'as* [Water in Strife and Action] (Israel: Massada Ltd., 1973), 125–128.
12. Ibid., 125.
13. Donna M. Herzog, "Contested Waterscapes: Constructing Israeli Water and Identity," PhD Dissertation, New York University, 2015, 70.

14. Blass, *op. cit.*, 129–130.
15. Elisha Kally and Gideon Fishelson, *Water and Peace: Water Resources and the Arab-Israeli Peace Process* (Westport, CT: Praeger, 1993), 6–7.
16. For a recounting of Lowdermilk's trip to Israel with the US Department of Agriculture, see: Walter Clay Lowdermilk, *Conquest of the Land through Seven Thousand Years* (Washington, DC: U.S. Department of Agriculture, 1948).
17. Walter Clay Lowdermilk, *Palestine, Land of Promise* (New York: Harper & Brothers, 1944), 5.
18. Ibid., 4.
19. Ibid., 148–161.
20. Harper & Brothers first released Walter Clay Lowdermilk's *Palestine, Land of Promise* in 1944.
21. Amir Mane, "Americans in Haifa: The Lowdermilks and the American-Israeli Relationship," *Journal of Israeli History* 30, no. 1 (2011), 71. *Palestine, Land of Promise* was also translated into seven languages, including Hebrew and Yiddish.
22. R. L. Duffus, "Practical View of Palestine," review of *Palestine, Land of Promise*, ed. Walter Clay Lowdermilk, *The New York Times,* May 21, 1944, Sunday Book Review.
23. Excerpt from the review of *Palestine, Land of Promise* in the *New York Herald Tribune*: "Mr. Lowdermilk gives a report, and a very exciting one, on what the Zionists are up to. It is exciting because Mr. Lowdermilk has a special point of view. He examines the experiment in Palestine, not as a Jew and with no more sympathy for the problems of the Jews than any other person of normal humanitarian instincts, but as a soil conservationist! . . . Mr. Lowdermilk's bold thesis is that a less halting support of the Palestine experiment—indeed, the wholehearted encouragement and expansion of this experiment—would, while giving scope to the aspiration for the Jews, be of great ultimate benefit to the Middle East, generally." [Philip Wagner, "The Miracle That Is Going on in Palestine: The Jews Restore Fertility Where the Desert Had Crept In," review of *Palestine, Land of* Promise, ed. Walter Clay Lowdermilk, the *New York Herald Tribune*, April 2, 1944, Weekly Book Review Section.]
24. Lowdermilk, *Palestine, Land of Promise*, *op. cit.*, 227.
25. Ibid.
26. Ibid., 229.
27. Guilford Glazer, author's interview, telephone, December 12, 2012.
28. Inez Marks Lowdermilk, *All in a Lifetime: An Autobiography* (Berkeley, CA: The Lowdermilk Trust, 1985), 229.
29. Ibid.
30. Westher Hess, author's interview, telephone, April 2, 2014.

31. Makovsky, *op. cit.*, 238.
32. The British position on Palestine in 1945: "It is indispensable to Imperial security in the Middle East . . . that Palestine should be administered by Britain as an undivided whole. . . . Palestine and Transjordan must constitute the core of our Middle Eastern system of security. . . . The Middle Eastern Defense Committee is unanimous in the opinion that the partition of Palestine would, from the military standpoint, spell irremediable disaster." [Minister Resident in the Middle East, "Imperial Security in the Middle East," July 2, 1945, 7.]
33. There were several approaches by the different Zionist groups as to how to challenge British rule. Ben-Gurion and the Labor Zionists generally favored negotiation and political means. The Irgun, or IZL, and the Stern Gang, or Lehi, chose more violent confrontation. For more information on the competing approaches, see: Howard M. Sachar, *A History of Israel: From the Rise of Zionism to Our Time* (New York: Knopf, 1976), 249–278.
34. For deeper insight into David Ben-Gurion's views on the Negev, see: David Ben-Gurion, "Introduction," in *Masters of the Desert: 6,000 Years in the Negev*, ed. Yaakov Morris (New York, G. P. Putnam's Sons, 1961), 11–16.
35. Kellerman, *op. cit.,* 248–249.
36. Blass, *op. cit.*, 141–143.
37. Uri Werber, author's interview, Kibbutz Hatzerim (Israel), May 5, 2013.
38. Blass, *op. cit.*, 142.
39. Ibid., 145.
40. Mekorot, *60 Shanah Le-Kav Ha-Rishon La-Negev* [60 Years to the First Pipeline to the Negev] (Tel Aviv: Mekorot, 2007).
41. The 1931 Census of Palestine recorded the population of the Beersheba region at 51,082, of which 47,981 were nomadic people and 3,101 were living in settled communities.* By 1948, the population was estimated at 70,000, the majority of whom were Bedouin.** Thus, the actual population in 1947 was somewhere between those two figures. Using the 1948 figure, with a total Negev landmass of 4,700 square miles, the Negev had a population density in 1948 of about fifteen people per square mile.

    * Government of Palestine, *Census of Palestine 1931, Volume II. Palestine, Part II, Tables*, ed. E. Mills (Alexandria: Government of Palestine, 1933), 2–3.

    ** Shlomo Swirski and Yael Hasson, *Invisible Citizens: Israel Government Policy toward the Negev Bedouin*, trans. Ruth Morris (Tel Aviv: Adva Center, 2006), 9.
42. Technion professor and hydrologist Uri Shamir described Blass's testimony as "Zionist hydrology," more ideology than engineering or science, even if Blass ended up being correct. [Uri Shamir, author's interview, Caesarea (Israel), May 1, 2013.]

43. For an analysis of the 1948 Arab-Israeli War, see: Benny Morris, *Righteous Victims: A History of the Zionist-Arab Conflict, 1881–2001* (New York: Vintage, 2001), 215–258.
44. For more information on the expulsion of the Jews from Arab countries following the creation of the State of Israel, see: Martin Gilbert, *In Ishmael's House: A History of Jews in Muslim Lands* (New Haven, CT: Yale University Press, 2010), 217–281.
45. Central Bureau of Statistics, *Israel in Statistics: 1948–2007* (Jerusalem: Central Bureau of Statistics, May 2009), 2.
46. Israel's first three-and-a-half years of immigration totals are: 1948 (101,828), 1949 (239,954), 1950 (170,563), and 1952 (175,279). [Ministry of Foreign Affairs, *Population of Israel: General Trends and Indicators* (Jerusalem: Ministry of Foreign Affairs, December 24, 1998).]
47. Nadav Morag, "Water, Geopolitics and State Building: The Case of Israel," *Middle Eastern Studies* 37, no. 3 (2001): 179–198.
48. Daniel Gordis, *Menachem Begin: The Battle for Israel's Soul* (New York: Schocken, 2014), 111.
49. Chris Sneddon and Coleen Fox, "The Cold War, the US Bureau of Reclamation, and the Technopolitics of River Basin Development, 1950–1970," *Political Geography* 30 (2011): 457.
50. W. H. Lawrence, "Eisenhower Sends Johnston to Mid-East to Ease Tension: Film Official Will Press for Israeli-Arab Accord and Economic Development," *The New York Times*, October 16, 1953.
51. For an account and analysis of "The Johnston Mission," see: Jeffrey Sosland, *Cooperating Rivals: The Riparian Politics of the Jordan River Basin* (Albany, NY: State University of New York Press, 2007), 37–61.
52. Blass, *op. cit.*, 203–204.
53. John S. Cotton, *Plan for the Development of the Water Resources of the Jordan and Litani River Basins* (Jerusalem: Ministry of Agriculture, 1954), 62.
54. For an analysis of the revisions to the Johnston Plan, see: Samer Alatout, "Hydro-Imaginaries and the Construction of the Political Geography of the Jordan River: The Johnston Mission, 1953–56," in *Environmental Imaginaries of the Middle East and North Africa*, eds. Diana K. Davis and Edmund Burke III (Athens, OH: Ohio University Press, 2011).
55. "Israel Inaugurates Yarkon-Negev Pipeline Amid Great Festivities," *Jewish Telegraphic Agency*, July 29, 1955.
56. Seltzer, *op. cit.*, 128.

57. Bezalel Amikam, "Ish Ha-Mayim" [The Water Man], *Al HaMishmar*, August 27, 1982.
58. David Ben-Gurion, *Letter to Simcha Blass*, March 3, 1956.
59. Yael Shoham and Ofra Sarig, *Ha-Movil Ha-Artzi: Min Ha-Kineret ve-ad Pe-atei Negev* [The National Water Carrier: From the Sea of Galilee to the Negev] (Tel Aviv: Mekorot, 1995), back cover page.
60. Noel Maurer and Carlos Yu, "What Roosevelt Took: The Economic Impact of the Panama Canal, 1903–37," Harvard Business School, Work Paper: 06-041 (March 2006), 3.
61. In 1961, two years after construction on the National Water Carrier had begun, the Beersheba metropolitan area population was 97,200. Today it is 664,000. [Central Bureau of Statistics, *Statistical Abstract of Israel 2014* (Jerusalem: Central Bureau of Statistics, 2014).]
62. The author is grateful to Daniel Hoffman for the ideas expressed in this paragraph.
63. Seltzer, *op cit.*, 132.
64. Aaron Wiener (1912–2007) was also a brilliant water engineer. In interviews with his daughter and son-in-law, Ruti* and Uzi Arad,** it was made clear that Wiener, a low-key person, found working for the often cantankerous Blass an ordeal.

    * Ruti Arad, author's interview, telephone, March 6, 2013.

    ** Uzi Arad, author's interview, New York, March 30, 2013.
65. "Hi'gia Professor Lowdermilk" [Professor Lowdermilk Arrives], *Davar*, June 7, 1964.

CHAPTER 3: MANAGING A NATIONAL WATER SYSTEM

1. State of Israel, *Water Law*, 5719-1959, Section 125–126(a).
2. Uri Shani, author's interview, telephone, March 17, 2013.
3. David Pargament, author's interview, Tel Aviv, April 26, 2013.
4. Yoav Kislev, *The Water Economy of Israel* (Jerusalem: Taub Center for Social Policy Studies in Israel, November 2011), 104.
5. Olga Slepner, e-mail to the author, April 23, 2014.
6. Oded Fixler, author's interview, Tel Aviv, May 6, 2013.
7. Shani, *op. cit.*
8. "Israel Spells Out 2010 Tariff Plan," Global Water Intelligence, November 19, 2009.
9. Nir Barlev, author's interview, telephone, April 9, 2013.
10. Ibid.
11. Controversy over the municipal utilities continues, with mayors and their political

allies attempting to transfer local control of water back to the mayors. [Avi Bar-Eli, "Be'lakhatz Ha-Reshuyot Ha-Mekomiyot—Lapid Hit'kapel Ve'Shina et Khok Ta'agidei Ha'Mayim" [Under Pressure from Local Authorities—Lapid Capitulates and Changes the Law on Water Corporations], *The Marker*, January 7, 2014.]

12. Shimon Tal, author's interview, telephone, March 6, 2013.
13. Nir Barlev, author's interview, telephone, April 11, 2013.
14. Taniv Rophe, author's interview, telephone, October 7, 2013.
15. Shimon Tal, *op. cit.*, March 6, 2013.
16. Ibid.
17. Some world cities with unaccounted for water losses at forty percent and above include: Delhi (53%), Dublin (40%), Glasgow (44%), Hyderabad (50%), Jakarta (51%), Montreal (40%), and Sofia (62%), among others. [Smart Water Networks Forum Research, *Stated NRW (Non-Revenue Water) Rates in Urban Networks* (Portsmouth, U.K.: Smart Water Networks Forum, August 2010).]
18. Olga Slepner, e-mail to the author, November 26, 2014.
19. Abraham Tenne, author's interview, Tel Aviv, April 25, 2013.
20. Slepner, *op. cit.*
21. Barlev, *op. cit.*, April 11, 2013.
22. Ibid.
23. Ibid.
24. Zohar Yinon, author's interview, Jerusalem, April 24, 2013.
25. Ibid.

CHAPTER 4: REVOLUTION(S) ON THE FARM

1. Simcha Blass, *Mei Meriva u-Ma'as* [Water in Strife and Action] (Israel: Massada Ltd., 1973), 330–331.
2. For a history of irrigation in the ancient Middle East, see: Sandra Postel, *Pillar of Sand: Can the Irrigation Miracle Last?* (New York: W. W. Norton & Company, 1999), 13–39.
3. Sandra Postel, "Drip Irrigation Expanding Worldwide," *News Watch*, June 25, 2012.
4. Ibid.
5. Ministry of Agriculture and Rural Development, *Irrigation Agriculture—The Israeli Experience*, ed. Anat Lowengart-Aycicegi (Jerusalem: Ministry of Agriculture), 6.
6. In 1962, Israel's agricultural sector consumed 78% of the total water supply. [Aaron Wiener, *Development and Management of Water Supplies under Conditions of Scarcity of Resources*. Tel Aviv: TAHAL, April 1964.]

7. United Nations World Water Assessment Program, *The United Nations World Water Development Report 2014: Water and Energy* (Paris: United Nations Education, Scientific and Cultural Organization, 2014), 56.
8. Netafim, *Irrigation and Strategies for Investment* (Presentation, Naty Barak, Agricultural Investment 2011, London, 5–6, October 2011).
9. Simcha Blass, *Drip Irrigation* (Tel Aviv: Water Works—Consulting and Design, July 1969), 3.
10. After seeing his academic career blocked, the faculty member, Dan Goldberg, became a consultant in Caribbean and South American countries in using drip irrigation for growing bananas.* Goldberg became a lifelong evangelist for drip irrigation, and coauthored a book on the subject.**

    * Yossi Shalhevet, author's interview, telephone, October 3, 2014.

    ** Dan Goldberg, Baruch Gornat, and D. Rimon, *Drip Irrigation: Principles, Design, and Agricultural Practices* (Kfar Shmaryahu, Israel: Drip Irrigation Scientific Publications, 1976).
11. Uri Werber, author's interview, Kibbutz Hatzerim (Israel), May 5, 2013.
12. Zwi Keren, *Oasis in the Desert: The Story of Kibbutz Hatzerim*, trans. Kfar Blum Translation (Kibbutz Hatzerim, Israel: Kibbutz Hatzerim Press, 1988), 159–164.
13. Werber, *op. cit.*
14. Ibid.
15. Daniel Gavron, *The Kibbutz: Awakening from Utopia* (Washington, DC: Rowman & Littlefield Publishers, 2000), 124–125.
16. Ruth Keren, author's interview, Kibbutz Hatzerim (Israel), May 5, 2013.
17. The three kibbutzim that set up drip-irrigation companies in the 1970s that competed with Netafim were: Kibbutz Gvat (Plastro), Kibbutz Na'an (Na'an), and Kibbutz Dan (Dan). Na'an and Dan merged in 2001 doing business thereafter as NaanDan. In addition, several others—mostly from kibbutzim—entered the drip-irrigation business albeit with less success. Of that second tier, Metzerplas of Kibbutz Metzer was the most successful. It is still in the drip-irrigation business.
18. Naty Barak, author's interview, New York, March 21, 2013.
19. Plastro was sold to John Deere and did business under the name John Deere Water until it was sold again to a private equity firm.* NaanDan is owned by Jain Irrigation, a large Indian company.**

    * Yoram Gabison, "FIMI wins auction for control of John Deere Water," *Haaretz*, February 17, 2014.

    ** "About Us," NaanDanJain, accessed on November 26, 2014: www.naandanjain.com/Company/Irrigation-Solutions/.

20. Werber, *op. cit.*
21. Erez Meltzer, author's interview, telephone, January 23, 2013.
22. Werber, *op. cit.*
23. Author's visit, May 5, 2013.
24. Barbara Shivek, author's interview, Kibbutz Hatzerim (Israel), May 5, 2013.
25. Except in special cases, military service is universal in Israel. After they turn eighteen, men serve for three years and women for two.
26. Rafi Mehoudar, author's interview, Tel Aviv, April 18, 2013.
27. Postel, "Drip Irrigation Expanding Worldwide," *op. cit.*
28. Netafim, *Drip Irrigation—Israeli Innovation That Has Changed the World* (Presentation, Naty Barak, JNF Summit, Las Vegas, April 28, 2013), 19.
29. Naty Barak, *op. cit.*
30. Mehoudar, *op. cit.*
31. Hebrew University Professor Aharon Friedman explains the plant physiology and resulting yield this way: "Plants do better when water is available because they can then afford to lose water by evaporation through the stomata while maintaining respiration and photosynthesis. In absence of water, the stomata remain open for shorter periods, hence respiration and CO_2 fixation is reduced—this is manifested by a slower growth rate and less yield." [Aharon Friedman, e-mail to the author, October 3, 2014.]
32. John Seewer, "Toledo, Ohio Water Contamination Leaves Residents Scrambling to Buy Bottled Water," *Huffington Post*, October 2, 2014.
33. Danny Ariel, author's interview, Tel Aviv, October 28, 2013.
34. Mehoudar, *op. cit.*
35. Uri Shani, author's interview, telephone, July 4, 2013.
36. Ibid.
37. Hazera Genetics, *Hazera—History of Success* (YouTube, 2011), accessed on March 9, 2015: www.youtube.com/watch?v=mKItOZwrzRY.
38. Shoshan Haran, author's interview, telephone, July 1, 2013.
39. Nili Shalev, e-mail to the author, September 5, 2014.
40. Zvi Amit, e-mail to the author, February 2, 2015.
41. Moshe Bar, author's interview, telephone, December 26, 2013.
42. Shoshan Haran, author's interview, telephone, July 1, 2013.
43. Zvi Amit, author's interview, telephone, July 10, 2013.
44. Blass, *Drip Irrigation*, *op. cit.*, 3.
45. Amit, interview, *op. cit.*
46. Bar, *op. cit.*

47. Haran, *op. cit.*
48. Ami Shaham, author's interview, Central Arava (Israel), April 23, 2013.
49. Arie Issar, author's interview, Jerusalem, April 24, 2013. For more information on the development of the water resources of the Negev's Central Arava, see: Government of Israel, *The Central Arava: Proposals for the Development of Water Resources*, Report 69-093 (Jerusalem: Government of Israel, September 1969) and Government of Israel, *The Central Arava: Irrigation Water Development Scheme*, Report 69-173 (Jerusalem: Government of Israel, November 1969).
50. Naty Barak, author's interview, telephone, November 7, 2013.
51. International Commission on Irrigation and Drainage, *Sprinkler and Micro Irrigated Areas* (New Delhi: International Commission on Irrigation and Drainage, May 2012).
52. International Commission on Irrigation and Drainage, *World Irrigated Area-Region Wise/Country Wise* (New Delhi: International Commission on Irrigation and Drainage, 2012).
53. International Commission on Irrigation and Drainage, *Sprinkler and Micro Irrigated Areas*, *op. cit.*
54. Fifteen percent of irrigated fields use sprinkler irrigation, but the use of flood irrigation is likely to continue for as long as farmers are given water at a cost close to zero. With no incentive to incur the cost of purchasing irrigation equipment, highly wasteful water practices remain in effect, even in places facing water stresses. For example, according to the Texas Water Resource Institute, the farmers in Texas only use drip irrigation on three percent of the state's crops. Until farmers have to factor in the real cost of water in their agriculture, this is likely to continue. [Todd Woody, "How Israel Beat a Record-Breaking Drought, With Water to Spare," *Takepart*, October 6, 2014.]
55. Shani, *op. cit.*
56. If California were a country, it would be one of the greatest users of drip irrigation. It now employs drip-irrigation systems on thirty-nine percent of its irrigated fields, including seventy-five percent of its vineyards. [Gwen N. Tindula, Morteza N. Orang, and Richard L. Snyder, "Survey of Irrigation Methods in California in 2010," *Journal of Irrigation and Drainage Engineering* 139, no. 3 (August 2013): 233–235.]
57. Postel, "Drip Irrigation Expanding Worldwide," *op. cit.*
58. Yuval Azulai, "Kibbutz Naan Sells NaanDanJain Irrigation," *Globes*, May 14, 2012.
59. Barak, *op. cit.*, March 18, 2013.
60. According to the World Bank: "Subsidies to utility customers are a salient feature of water and electricity services worldwide. In some cases, subsidized service is made possible by large transfers from general tax revenue, which can be in the form of either capital projects

or regular transfers to cover revenue shortfalls. . . . Other utilities simply absorb the financial loss from the general or targeted subsidies, gradually wearing down capital stock and pushing repair and maintenance costs off into the future." [World Bank, *Water, Electricity, and the Poor: Who Benefits from Utility Subsidies?* eds. Kristin Komives, Vivien Foster, Jonathan Halpern, and Quentin Wodon (Washington, DC: World Bank, 2005), 1.]

61. Uri Shamir, *Management of Water Systems under Uncertainty* (Talk, WATEC Conference, Tel Aviv, October 22, 2013).
62. There is universal agreement in India that drip irrigation is among the best tools for controlling water use, enhancing yield, and addressing poverty,* but there is dispute on the efficacy of the subsidies for drip irrigation and how to achieve the best economic outcome.**

    * Government of India, Ministry of Agriculture, *Report on the Task Force on Microirrigation* (New Delhi: Ministry of Agriculture, January 2004), vii xxix.

    ** Hemant K. Pullabhotla, Chandan Kumar, and Shilp Verma, *Micro-Irrigation Subsidies in Gujarat and Andhra Pradesh: Implications for Market Dynamics and Growth* (Sri Lanka: IWMI-TATA Water Policy Program, 2012). See also: Ravinder P. S. Malik and M. S. Rathore, *Accelerating Adoption of Drip Irrigation in Madhya Pradesh, India, AgWater Solutions Project* (Sri Lanka: IWMI, September 2012), 15–32.
63. Archana Chaudhary, "Netafim to Build Largest India's Drip-Irrigation Project," *Bloomberg*, January 23, 2014.
64. Barak, *op. cit.*, March 18, 2013.

CHAPTER 5: TURNING WASTE INTO WATER

1. Hillel I. Shuval, *Public Health Aspects of Waste Water Utilization in Israel* (Presentation, The Purdue Industrial Wastes Conference, May 1, 1962).
2. Ibid., 4.
3. Avi Aharoni, author's interview, Tel Aviv, January 6, 2014.
4. For a superb telling of the birth of municipal sewage treatment, see: Steven Solomon, *Water: The Epic Struggle for Wealth, Power and Civilization* (New York: Harper Perennial, 2010), 249–265. See also: James Salzman, *Drinking Water: A History* (New York: Overlook Duckworth, 2010), 85–97.
5. Eytan Levy, author's interview, telephone, March 21, 2013.
6. Joanne E. Drinan, *Water & Wastewater Treatment: A Guide for the Nonengineering Professional* (Boca Raton, FL: CRC Press, 2001), 159–168.
7. Ibid., 169–173.
8. Ibid., 175–204.

9. Ibid., 207–220.
10. Adam Kanarek, author's interview, Tel Aviv, October 18, 2013.
11. Ibid.
12. Shuval, *op. cit.,* 9.
13. Ibid., 7.
14. The Shafdan facility treats and percolates approximately 95 million gallons of treated wastewater every day. [Nelly Icekson-Tal, author's interview, Rishon LeZion (Israel), October 17, 2013.]
15. Ibid.
16. Moshe Gablinger, author's interview, telephone, April 21, 2014.
17. Ori Yogev, author's interview, Tel Aviv, April 19, 2013.
18. Aharoni, *op. cit.*
19. Shuval, *op. cit.,* 4.
20. Mekorot, *Wastewater Reclamation and Reuse*, eds. Batya Yadin, Adam Kanarek, and Yael Shoham (Tel Aviv: Mekorot, 1993), 9–14.
21. While waiting for the pipeline to the Negev to be completed, the Ministry of Health became convinced that Shafdan's SAT water was safe to drink in limited quantities. For most of the 1980s, the ministry permitted up to five percent of the nation's drinking water to be provided by Shafdan's SAT water. By 1989, when the Negev pipeline opened, the ministry decided that it would be best for treated wastewater and freshwater to be separated, and the practice came to an end. [Israel Mantel, author's interview, Tel Aviv, May 6, 2013.]
22. Aharoni, *op. cit.*
23. Taniv Rophe, author's interview, telephone, October 7, 2013.
24. Aharoni, *op. cit.*
25. Ibid.
26. In 2010, Israel's agricultural exports totaled $2.1 billion. [Ministry of Agriculture, "Israel's Agriculture at a Glance," ed. Arie Regev, in *Israel's Agriculture*, ed. The Israel Export and International Cooperation Institute (Tel Aviv: The Israel Export and International Cooperation Institute, 2012), 8.]
27. Shimon Tal, author's interview, Tel Aviv, October 18, 2013.
28. Avi Aharoni, e-mail to the author, October 5, 2014.
29. Yossi Schreiber, author's interview, telephone, September 21, 2014.
30. As Israel began building out its wastewater infrastructure, a large number of emigrants departing the former Soviet Union were resettled in Israel. Although not provided exclusively for the national wastewater infrastructure, the US provided loan

guarantees to assist in the immigrant absorption with the understanding that some of those loan guarantees would be used in helping to arrange lower-cost financing for the wastewater system. Israel received lower-rate financing at what was ultimately no cost to the US taxpayer as the loan guarantees were never invoked by any of the lenders.

31. Schreiber, *op. cit.*
32. Ibid.
33. Rophe, *op. cit.*
34. Ibid.
35. Uri Schor, author's interview, Tel Aviv, April 25, 2013.
36. Olga Slepner, e-mail to the author, November 27, 2014.
37. Shaul Ashkenazy, author's interview, telephone, October 6, 2013.
38. Ibid.
39. Slepner, *op. cit.*
40. Aharoni, *op. cit.*
41. Electric Power Research Institute, *Water and Sustainability (Volume 4): U.S. Electricity Consumption for Water Supply and Treatment—The Next Half Century* (Palo Alto, CA: Electric Power Research Institute, March 2002), vi.
42. Levy, *op. cit.*
43. Noah Galil, author's interview, Haifa, January 7, 2014.
44. Ibid.
45. Refael Aharon, e-mail to the author, February 5, 2015.
46. Although there is a high volume of salt in wastewater due to salt added in the cooking process everywhere, it may be even higher in Israel. The Jewish dietary laws include a ritual of salting all meat before cooking. That salt is rinsed off, but it flows into Israel's wastewater.
47. Dan Zaslavsky, author's interview, Haifa, January 7, 2014.
48. Steven Mithen, *Thirst: Water and Power in the Ancient World* (Cambridge, MA: Harvard University Press), 63.
49. Ibid., 44–74.
50. Dror Avisar, author's interview, Tel Aviv, January 6, 2014.
51. Ibid.
52. Sara Elhanany, author's interview, Tel Aviv, April 25, 2013.
53. Avisor, *op. cit.*
54. Aly Thomson, "Birth Control Pill Threatens Fish Populations," *The Canadian Press*, October 13, 2014.
55. Avisar, *op. cit.*

56. Elhanany, *op. cit.*
57. Oren Blonder, e-mail to the author, October 5, 2014.
58. Elhanany, *op. cit.*
59. Efi Stenzler, author's interview, New York, February 1, 2013.
60. For an analysis on the global impact of desertification, see: Anton Imeson, *Desertification, Land Degradation, and Sustainability* (Hoboken, NJ: Wiley, 2012).
61. Stenzler, *op. cit.*
62. Rophe, *op. cit.*
63. Sharon Udasin, "Israel, Greek, Cypriot Environment Ministries to Cooperate on Mediterranean Pollution Prevention," *The Jerusalem Post*, May 14, 2014.

CHAPTER 6: DESALINATION: SCIENCE, ENGINEERING, AND ALCHEMY

1. "Weizmann Institute Erects Plant in Palestine to Desalt for Drinking Purposes," *Jewish Telegraphic Agency,* March 1, 1948.
2. Ora Kedem, author's interview, telephone, December, 17, 2013.
3. Shimon Peres, author's interview, Tel Aviv, April 25, 2013.
4. Lyndon B. Johnson, "If We Could Take the Salt Out of Water," *The New York Times Magazine*, October 30, 1960.
5. James D. Birkett, "A Brief Illustrated History of Desalination: From the Bible to 1940," *Desalination*, no. 50 (1984): 17.
6. Lyndon B. Johnson, *op. cit.*
7. Charles F. MacGowan, *History, Function, and Program of the Office of Saline Water* (Presentation, New Mexico Water Conference, July 1–3, 1963), 24–33.
8. For one example of Lyndon Johnson's support for water bills, especially when a desalination component was added, see the *Demonstration Plants Act of 1958, Public Law 85-883*, initially introduced by Senator Clinton Anderson of New Mexico.
9. Lyndon B. Johnson, *Remarks in New York City at the Dinner of the Weizmann Institute of Science* (New York, February 6, 1964).
10. Dana Adams Schmidt, "Johnson Speech Infuriates Arabs: They Attack Offer to Help Israel Utilize Sea Water," *The New York Times*, February 8, 1964.
11. Memorandum of telephone conversation between Robert W. Komer, National Security Council Staff and George W. Ball, Under Secretary of State (Washington, DC, June 2, 1964), Foreign Relations of the United States, 1964–1968, Volume XVIII, Arab-Israeli Dispute, 1964–67, Document 66.
12. Peres, *op. cit.*

13. Nathan Berkman, author's interview, Tel Aviv, October 25, 2013.
14. Ibid.
15. See, for example, Ben-Gurion diary entries on August 16, 1954, August 20, 1954, February 7, 1956, June 11, 1957, and April 8, 1961, among many others.
16. "Israel to Remove Sea Water Brine," *The New York Times*, November 9, 1958.
17. "Science: Salt Water into Fresh," *Time*, September 3, 1956.
18. David Ben-Gurion, *Diary*, August 16, 1954.
19. For a lengthy contemporary account of Zarchin and his theory, see: Yaakov Morris, *Masters of the Desert: 6000 Years in the Negev* (New York: G. P. Putnam's Sons, 1961), 240–252.
20. David Ben-Gurion, *Diary*, September 30, 1955. David Ben-Gurion also diarized about the costs of implementing either the Zarchin plan for desalinating water (freezing) or the possibility of purifying seawater via heating pools in various stages. [David Ben-Gurion, *Diary*, February 7, 1956.]
21. Yosef Almogi, *Total Commitment* (East Brunswick, NJ: Cornwall Books, 1982), 198–199.
22. Alexander Zarchin was so adamant that his process was superior to other plans being considered by the government that he published a lengthy article in *Haaretz* on November 20, 1963, entitled "Purifying Water Instead of the National Water Carrier." The article was a sharp critique of the National Water Carrier in which he reported writing letters to government ministers in opposition to the national water system being constructed. Zarchin proposed that the National Water Carrier be scrapped and in its place plants for purifying seawater should be established in the Negev, citing issues of relative cost, salinity, and environmental degradation. To a degree, his approach won the day—only decades later.
23. Berkman, *op. cit.*
24. Ibid.
25. Avi Kay, "From *Altneuland* to the New Promised Land: A Study of the Evolution and Americanization of the Israeli Economy," *Jewish Political Studies Review* 24, no. 1–2 (2012): 103.
26. Editorial Note, Foreign Relations of the United States, 1964–1968, Volume XXXIV, Energy Diplomacy and Global Issues, Document 130. Also, in the author's interview with Miriam Eshkol, Levi Eshkol's widow, she said that her husband would often say that "water to the country is like blood to a human being." [Miriam Eshkol, author's interview, telephone, April 29, 2013.]
27. Memorandum of Conversation (Washington, DC, June 1, 1964), Foreign Relations of

the United States, 1964–1968, Volume XVIII, Arab-Israeli Dispute, 1964–67, Document 65.

28. Ibid.
29. Memorandum from Robert W. Komer, National Security Council Staff, to President Lyndon Johnson (Washington, DC, May 28, 1964), Foreign Relations of the United States, 1964–1968, Volume XVIII, Arab-Israeli Dispute, 1964–67, Document 63.
30. MacGowan, *op. cit.*
31. Lyndon B. Johnson, *Diary*, June 2, 1964.
32. Letter from Dean Rusk, Secretary of State, to Glenn Seaborg, Chairman of the Atomic Energy Commission (Washington, DC, December 9, 1964), Foreign Relations of the United States, 1964–1968, Volume XXXIV, Energy Diplomacy and Global Issues, Document 136.
33. Editorial Note, Foreign Relations of the United States, 1964–1968, Volume XXXIV, Energy Diplomacy and Global Issues, Document 149.
34. Editorial Note, Foreign Relations of the United States, 1964–1968, Volume XXXIV, Energy Diplomacy and Global Issues, Document 151.
35. "Bunker and Eshkol Confer About Desalting Plant," *The New York Times*, December 19, 1966.
36. Memorandum from Walt Rostow, Special Assistant to President Lyndon Johnson (Washington, DC, January 5, 1968), Foreign Relations of the United States, 1964–1968, Volume XX, Arab-Israeli Dispute, 1967–68, Document 33.
37. Memorandum of Conversation (LBJ Ranch, Texas, January 8, 1968, Session III), Foreign Relations of the United States, 1964–1968, Volume XX, Arab-Israeli Dispute, 1967–68, Document 41.
38. Memorandum from Henry Kissinger, Assistant for National Security Affairs to President Richard Nixon (Washington, DC, February 10, 1969), Foreign Relations of the United States, 1969–1976, Volume XXIV, Middle East Region and Arabian Peninsula, 1969–1972; Jordan, September 1970, Document 4.
39. National Archives, Nixon Presidential Materials, NSC Files, NSC Institutional Files (H-Files), Box H-141, National Security Study Memoranda, NSSM 30.
40. Memorandum from the Theodore Eliot, Executive Secretary of the Department of State, to Henry Kissinger, President's Assistant for National Security Affairs (Washington, DC, December 6, 1972), Foreign Relations of the United States, 1969–1976, Volume XXIV, Middle East Region and Arabian Peninsula, 1969–1972; Jordan, September 1970, Document 35.
41. In the author's interview with Nathan Berkman, he reported having no memory of

Johnson's Weizmann Institute speech, but said that when Eshkol returned to Israel in 1964 everyone in the Israel desalination community was focused on how to develop a program that would qualify for the funding. Berkman said that there had been a joke circulating in his department after the 1964 Eshkol visit to Washington, DC, that the US would provide $100 million to provide the facility and that Israel would provide the seawater. Despite the joke, he said, everyone knew that Israel would only get the funding if it came up with a breakthrough idea.

42. Berkman, *op. cit.*
43. Nathan Berkman, "Back in the Old Days," *Israel Desalination Society*, 2007.
44. "U.S., Israel Finally to Build Horizontal Tube Prototype at Ashdod," *Water Desalination Report* 11, no. 27 (July 3, 1975).
45. Jeremy Sharp, *U.S. Foreign Aid to Israel* (Washington, DC: Library of Congress, Congressional Research Service, April 11, 2014), 28.
46. Memorandum from Eliot to Kissinger, *op. cit.*
47. Berkman, interview, *op. cit.*
48. IDE was merged with Israel Chemicals in the 1980s.* Israel Chemicals, which is part of Israel Corporation, was sold to the Ofer Brothers Group in 1999.** Delek Group bought 50 percent of IDE in 2000.*

    * IDE Technologies Ltd., *IDE Technologies Limited* (Presentation, Gal Zohar, CFO Forum, 2011).

    ** Orna Raviv, "Israel Corp Sale Completed; Ofer Family Expected to Restructure Group," *Globes*, April 15, 1999.
49. Fredi Lokiec, author's interview, Kadima (Israel), May 1, 2013.
50. "Carlsbad Desalination Plant to Utilize IDE Technologies' Reverse Osmosis Solution," *Water World*, January 2, 2013.
51. "Toanjin SDIC Project: China's Largest Desalination Plant," IDE Technologies, accessed on February 4, 2015: www.ide-tech.com/blog/case-study/tianjin-china-project-ide/.
52. "Gujarat Reliance Project: India's Largest Desalination Plant," IDE Technologies, accessed on February 4, 2015: www.ide-tech.com/blog/case-study/reliance-project-2/.
53. "Sorek Project: The World's Largest and Most Advanced SWRO Desalination Plant," IDE Technologies, accessed on February 4, 2015: www.ide-tech.com/blog/case-study/sorek-israel-project/.
54. Ronen Wolfman, author's interview, telephone, February 20, 2014.
55. Ibid.
56. Rafi Semiat, author's interview, Haifa, May 2, 2013.

57. Ibid.
58. In the author's interview with Tami Schor, a senior official at the Israel Water Authority, she pointed out that there are always short-term fixes available to boost water production like opening closed wells, adding a level of (expensive) purification to contaminated water sources, and so forth. Her point was that it isn't a sustainable approach, even if it might yield positive short-term benefits. [Tami Schor, author's interview, Tel Aviv, January 6, 2014.]
59. Ilan Cohen, author's interview, telephone, March 29, 2013.
60. Avraham Baiga Shochat, author's interview, Tel Aviv, January 8, 2014.
61. Alan Philps, "Drought Forces Israel to Import Turkish Water," *The Telegraph*, June 28, 2000.
62. Ram Aviram, author's interview, New York, February 7, 2013.
63. Uri Shani, author's interview, telephone, March 17, 2013.
64. Ronen Wolfman, author's interview, Ramat Gan, Israel, October 24, 2013.
65. Ibid.
66. Miriam Balaban, author's interview, telephone, October 3, 2013.
67. Coalinga's name is a vestige of its first function. When the California railway system was constructed in the 1880s, all of the trains ran on coal. Because Coalinga was the first of the coaling stations where the steam engines would take on coal to power trains, the place was named Coaling A, down the line from Coaling B and Coaling C. When the railroad switched to oil, the tiny town lost its relevance, but it already seemed to have a name. Coaling A became Coalinga.
68. Mickey Loeb, author's interview, Omer (Israel), January 16, 2014.
69. Eilon Adar, author's interview, Beersheba (Israel), April 21, 2013.
70. Loeb, *op. cit.*
71. Berkman interview, *op. cit.*
72. Loeb, *op. cit.*
73. Tom Pankratz, author's interview, telephone, August 12, 2014.
74. "Market Data, Technologies Used," Water Desalination Report, accessed on March 30, 2015: www.desalination.com/market/technologies.
75. Fredi Lokiec, *South Israel 100 Million m^3/Year Seawater Desalination Facility: Build, Operate and Transfer (BOT) Project* (Kadima, Israel: IDE Technologies Ltd., March 2006). By 2015 the Ashkelon desalination plant had produced more than 250 billion gallons of clean water, setting a world record. ["IDE's Israel Seawater RO Desalination Plant Sets World Record for Water Production," *WaterWorld*, February 10, 2015.]
76. Tom Pankratz, a desalination industry veteran, said: "Where Israel is extraordinary in

desalination is that Israel has built plants that are consistent, reliable, and produce high-quality water. They have made brilliant use of the variability of electricity tariffs and the desalination plants produce more at night and at low-peak periods to make use of that electrical variability. No one has come close anywhere in the world to using off-peak electricity the way Israel has. This may sound easy, but it requires an amazing balancing act to keep plants operating and to plan out when you will be accessing the electricity." [Pankratz, *op. cit.*]

77. Abraham Tenne, e-mail to the author, July 30, 2014.
78. Shimon Tal, author's interview, Tel Aviv, April 18, 2013.
79. Ilan Cohen, *op. cit.*
80. Abraham Tenne, Daniel Hoffman, and Eytan Levi, "Quantifying the Actual Benefits of Large-Scale Seawater Desalination in Israel," *Desalination and Water Treatment* 51, 1–3 (July 2012), 26–37.
81. Daniel Hoffman, an expert on desalination, writes: "I think that the main health benefit from the blending of high-quality desalinated water with water from natural sources (aquifers and Sea of Galilee) will be not in reducing chlorides and sodium, but in reducing the levels of nitrates and industrial contaminants that are currently present in the aquifers (and continuing to increase due to human activities above the aquifers). The current levels of nitrates in most of the wells in the Coastal Aquifer, for example, are above the European limit of 45–50 ppm, and certainly above the US limit which is 10 ppm. The Israeli standard for drinking water is currently at 70 ppm. High levels of nitrates are a danger to pregnant women, and could cause what is called 'blue baby syndrome.'" [Daniel Hoffman, e-mail to the author, August 16, 2014].
82. Tenne, Hoffman, and Levi, *op. cit.*, 26–27.
83. See: Introduction, note 31.
84. "Human Settlements on the Coast: The Ever More Popular Coasts," UN Atlas of the Oceans, accessed on March 25, 2015: www.oceansatlas.org/servlet/CDSServlet?status=ND0xODc3JjY9ZW4mMzM9KiYzNz1rb3M~.
85. Cohen, *op. cit.*

CHAPTER 7: RENEWING THE WATER OF ISRAEL

1. "History," Maccabiah, accessed on February 4, 2014: www.maccabiah.com/master/know-us-history.
2. Chuck Slater, "First-Hand Report of Maccabiah Tragedy," *The New York Times*, August 3, 1997.

3. Serge Schmemann, "2 Die at Games in Israel as Bridge Collapses," *The New York Times*, July 15, 1997.
4. Parliamentary [Knesset] Commission of Inquiry with Regard to Lessons to Be Learned from the Maccabiah Bridge Disaster. *Report* (Jerusalem: Knesset: July 9, 2000).
5. "Death Tied to Pollution," *The New York Times*, July 28, 1997.
6. Serge Schmemann, "Israelis Turn Self-Critical as Mishap Kills Two," *The New York Times*, July 18, 1997.
7. David Pargament, e-mail to the author, September 10, 2014.
8. There are many examples of reverence for the land and its creatures in the Hebrew Bible. For a few examples note: Leviticus 25:23–24; Isaiah 24:4–6; Isaiah 43:20–21; Jeremiah 2:7; Ezekiel 34:2–4; Psalms 24:1; and Psalms 96:10–13.
9. See A. D. Gordon's *Our Tasks Ahead* (1920) in which the Zionist pioneer calls for a return to nature and labor in the soil of the Jewish homeland: "The Jewish people has been completely cut off from nature and imprisoned within city walls for two thousand years. . . . We lack the habit of labor . . . for it is labor which binds a people to its soil and to its national culture." [A. D. Gordon, "Our Tasks Ahead," in *The Zionist Idea: A Historical Analysis and Reader*, ed. Arthur Hertzberg (Philadelphia: Jewish Publication Society, 1997).]
10. John Stemple, "Viewpoint: A Brief Review of Afforestation Efforts in Israel," *Rangelands* 20, no. 2 (April 1998): 15–18.
11. Water Law, 5719-1959.
12. Streams and Springs Authorities Law, 5724-1965. In addition, rivers were addressed in the Drainage and Flood Control Law, 5718-1957, even if that was not specifically focused on the ecology of the country's rivers.
13. Shoshana Gabay, "Restoring Israel's Rivers," *Ministry of Foreign Affairs*, 2001.
14. Mekorot, *Masskinot Veh-ah-dat Ha-Yarkon* [Conclusions of the Committee on the Yarkon], ed. Simcha Blass (Tel Aviv: Mekorot, March 18, 1954). A coastal river is one that runs to the sea. Israel's longest river is the Jordan River, which terminates in the Dead Sea.
15. Alon Tal, *Pollution in a Promised Land: An Environmental History of Israel* (Berkeley, CA: University of California Press, 2002), 8–9.
16. David Pargament, author's interview, Tel Aviv, April 26, 2013.
17. David Pargament, e-mail to the author, March 14, 2015.
18. Ibid.
19. David Pargament, e-mail to the author, November 22, 2014.
20. Pargament, interview, *op. cit.*
21. Ibid.

22. The full name of the river is the Hebron-Besor-Beersheba River. At different parts of the river, it takes on the name of one of those three locations, but the three rivers are all part of the same river system.
23. Richard Laster and Dan Livney, "Basin Management in the Context of Israel and the Palestinian Authority," in *Water Policy in Israel: Context, Issues and Options*, ed. Nir Becker (Dordrecht, Netherlands: Springer, 2013), 232.
24. Nechemya Shahaf, author's interview, Beersheba (Israel), April 21, 2013.
25. Ibid.
26. Jewish National Fund, *Blueprint Negev—Business Plan* (New York: Unpublished, 2004).
27. Russell Robinson, author's interview, New York, March 9, 2013.
28. Itai Freeman, author's interview, telephone, September 10, 2014.
29. Ibid.
30. Israel State Comptroller, *Report for 2011* (Jerusalem: State Comptroller, December 2011).
31. For an example of one such piece of legislation, see: "The Food Quality Protection Act Background," U.S. Environmental Protection Agency, accessed on December 1, 2014: www.epa.gov/pesticides/regulating/laws/fqpa/backgrnd.htm.
32. Israel State Comptroller, *op. cit.*
33. Eytan Israeli, author's interview, Kibbutz Kfar Blum (Israel), April 29, 2013.
34. For more information on "The Johnston Mission," see: Jeffrey Sosland, *Cooperating Rivals: The Riparian Politics of the Jordan River Basin* (Albany: State University of New York Press, 2007), 37–61.
35. Jeffery Sosland, author's interview, telephone, December 10, 2013.
36. Sosland, *Cooperating Rivals*, *op. cit.*, 179.
37. Ram Aviram, author's interview, New York, February 7, 2014.
38. Ibid.
39. Shimon Tal, author's interview, Tel Aviv, October 18, 2013.
40. Mike Rogoff, "The Ancient Galilee Boat," *Haaretz*, December 19, 2012.
41. Diego Berger and Meir Rom, author's interview, Sapir Pumping Station (Israel), April 29, 2013.
42. One concern likely unique to Israel that was recently revealed by the monitoring of the lake was the presence of nonnative microscopic snails. Although they were no threat to the health of the lake or to those who might drink the water pumped from the Sea of Galilee, shellfish is a forbidden food to those who observe the Jewish dietary laws. Special nonnative fish were introduced to the lake to eat the tiny snails so that Israel's religiously observant Jews would have no qualms about drinking tap

water. [Bonnie Azoulay, author's interview, Eshkol Filtration Plant (Israel), April 30, 2013.]

43. Yossi Shamaya, author's interview, Eshkol Filtration Plant (Israel), April 30, 2013.
44. Azoulay, *op. cit.*
45. The rain-cloud seeding process was invented in the US by writer Kurt Vonnegut's older brother Bernard and was the basis for the younger brother's apocalyptic novel *Cat's Cradle*. Silver iodide was memorably renamed "Ice-nine" in the novel. [Wolfgang Saxon, "Bernard Vonnegut, 82, Physicist Who Coaxed Rain from the Sky," *The New York Times*, April 27, 1999.]
46. Nati Glick, author's interview, telephone, June 13, 2013.
47. Dan Zaslavsky, author's interview, Haifa, January 7, 2014.
48. Joshua Schwarz, author's interview, telephone, October 7, 2014.
49. Azoulay, *op. cit.*
50. Tal, *op. cit.*

CHAPTER 8: TURNING WATER INTO A GLOBAL BUSINESS

1. Oded Distel, author's interview, New York, March 7, 2013.
2. Booky Oren, author's interview, New York, March 20, 2013. In addition to Oren, Ilan Cohen was the other wunderkind that Professor Dov Pekelman had asked to join his consulting company.
3. Dalia Tal, "Netafim VP Baruch Oren to Be Appointed Mekorot Chairman," *Globes*, September 7, 2003.
4. Oren, *op. cit.*
5. "Frequently Asked Questions: Pricing Water Services," U.S. Environmental Protection Agency, accessed on June 10, 2014: water.epa.gov/infrastructure/sustain/pricing_faqs.cfm.
6. Oren, *op. cit.*
7. Fanny Gidor, *Socio-Economic Disparities in Israel* (Piscataway, NJ: Transaction Publishers, 1979), 52.
8. Ministry of Agriculture, "Israel's Agriculture at a Glance," ed. Arie Regev, in *Israel's Agriculture*, ed. The Israel Export and International Cooperation Institute (Tel Aviv: The Israel Export and International Cooperation Institute, 2012), 8.
9. "Industrial Palestine," *The Economist*, August 15, 1942.
10. Paul Rivlin, *The Israeli Economy from the Foundation of the State through the 21st Century* (Cambridge: Cambridge University Press, 2011), 19.
11. Jacob Metzer, *The Divided Economy of Mandatory Palestine* (Cambridge: Cambridge University Press, 1998), 122.

12. Israel's population on May 15, 1948, was 806,000.* Its population in 1952 was 1,630,000.**

    * Central Bureau of Statistics, *Israel in Statistics: 1948–2007* (Jerusalem: Central Bureau of Statistics, May 2009), 2.

    ** Israel Ministry of Foreign Affairs, *Population of Israel: General Trends and Indicators* (Jerusalem: Ministry of Foreign Affairs, December 24, 1998).

13. For more information on the role of West German reparations, foreign aid, and donations in the Israeli economy, see: Bruce Bartlett, "The Crisis of Socialism in Israel," *Orbis* 35 (1991): 53–61.
14. Alan Dowty, "Israel's First Decade: Building a Civic State," in *Israel: The First Decade of Independence*, eds. Selwyn K. Troen and Noah Lucas (Albany: State University of New York Press, 1995), 46.
15. Mark Tolts, *Post-Soviet Aliyah and Jewish Demographic Transformation* (Presentation, Fifteenth World Congress of Jewish Studies, 2009), 3.
16. Ibid., 14–15.
17. Dan Senor and Saul Singer, *Start-Up Nation: The Story of Israel's Economic Miracle* (New York: Twelve, 2009).
18. According to the World Bank, Israel spent 5.6 percent of its GDP on defense in 2013—more than any other OECD country.* In 2009, according to a study by the Central Bureau of Statistics, Israel spent 18.7 percent of its national budget on defense—three times more than Britain and five times more than Germany.**

    * World Bank, *Military Expenditures (% of GDP)*.

    ** Moti Bassok, "Israel Shells Out Almost a Fifth of National Budget on Defense, Figures Show," *Haaretz*, February 14, 2013.

19. Organization for Economic Co-operation and Development, *Main Science and Technology Indicators 2014*, Issue 2 (Paris: OECD Publishing, 2015).
20. Inbal Orpaz, "R&D Culture: Israeli Enterprise, Chinese Harmony," *Haaretz*, January 7, 2014.
21. David Waxman, author's interview, telephone, February 21, 2013.
22. Lux Research, *Making Money in the Water Industry*, LRWI-R-13-4 (New York: Lux Research, December 2013), 2.
23. Waxman, *op. cit.*
24. Oren, *op. cit.*
25. Ori Yogev, author's interview, Tel Aviv, April 19, 2013.
26. Ilan Cohen, author's interview, telephone, March 29 2013; Oren, *op. cit.*; and Yogev, *op. cit.*
27. It was a prestigious group that joined in the effort with Waterfronts. These included (titles

are from around the time of 2005): Avner Adin, professor at Hebrew University; Ilan Cohen, director general of the Prime Minister's Office; Raanan Dinur, director of the Ministry of Commerce and Industry; Kalman Kaufman, venture partner at Israel Seed Partners; Booky Oren, chairman of Mekorot; Mira Rashty, COO of WhiteWater; Bob Rosenbaum, a marketing consultant; David Waxman, former CEO of IDE; Dan Wilensky, founder of Applied Materials; and Ori Yogev, chairman of WhiteWater.

28. Mira Rashty, author's interview, telephone, May 9, 2013.
29. Trigger Consulting, *Seizing Israel's Opportunities in the Global Water Market*, ed. Noam Gonen (Tel Aviv: Trigger Consulting, 2005).
30. Noam Gonen, author's interview, telephone, April 2, 2013.
31. "Gov't Launches New Water R&D Program," *Globes*, October 30, 2007.
32. Ibid.
33. As Arthur Ruppin, a leader of the *Yishuv* said: "The question was not whether group settlement was preferable to individual settlement; it was rather one of either group settlement or no settlement at all." [Paula Rayman, *The Kibbutz Community and Nation Building* (Princeton, NJ: Princeton University Press, 1981), 12.]
34. Michael Palgi and Shulamit Reinharz, *One Hundred Years of Kibbutz Life: A Century of Crises and Reinvention* (New Brunswick, NJ: Transaction, 2011), 2.
35. Gabriel Kahaner, *History of the Amiad Factory—In the Words of the Founder*, December 18, 2007.
36. Amiad Water Systems Limited, *Results for the Twelve Months to 31 December 2013*, 8.
37. Ibid.
38. Ran Israeli, author's interview, Tel Aviv, October 23, 2013.
39. Ibid.
40. Amos Shalev, author's interview, Tel Aviv, October 24, 2013.
41. Ibid.
42. Ariel Sagi, e-mail to the author, August 4, 2013.
43. "Company Profile of the Plasson Group," Plasson, Accessed on November 26, 2014, 2014: www.plasson.com/content/page/Profile-Plasson-group.
44. Booky Oren, e-mail to the author, November 10, 2014.
45. Rotem Arad, author's interview, Tel Aviv, October 24, 2013.
46. Ibid.
47. The sale price of Peleg's company, YaData, was not disclosed, but Microsoft was understood to have bought it for tens of millions of dollars. Amir Peleg owned 60 percent of the company. [Guy Grimland, "Microsoft Buys Startup YaData," *Haaretz*, February 28, 2008.]

48. Amir Peleg, author's interview, Tel Aviv, October 23, 2013.
49. Elie Ofek and Matthew Preble, "TaKaDu," Harvard Business School, Case Study 514-083 (January 2014).
50. Joshua Yeres, author's interview, Jerusalem, April 24, 2013.
51. Zohar Yinon, author's interview, Jerusalem, April 24, 2013.
52. Ofek, *op. cit.*
53. David Benovadia, "Using Water to Power Itself," *ISRAEL21c*, January 16, 2012.
54. Ministry of Economy, Office of the Chief Scientist, *Research and Development 2012–14* (Jerusalem: Office of the Chief Scientist, September 2014).
55. Ibid.
56. Yossi Smoler, author's interview, telephone, March 18, 2014.
57. Ibid.
58. Ministry of Economy, Office of the Chief Scientist, *Technological Incubator's Program*, ed. Yossi Smoler (Presentation, October 17, 2010).
59. Yossi Yaacoby, author's interview, Tel Aviv, May 6, 2013.
60. Oren Blonder, author's interview, telephone, March 25, 2014.
61. Yaacoby, *op. cit.*
62. Adi Yefet, e-mail to the author, March 18, 2014.
63. Booky Oren, interview, *op. cit.*
64. Distel, *op. cit.*
65. Cohen, *op. cit.*

CHAPTER 9: ISRAEL, JORDAN, AND THE PALESTINIANS: FINDING A REGIONAL WATER SOLUTION

1. See: Introduction, note 31.
2. According to Mekorot, "The cost to Mekorot of potable water provisions (including extraction, purification, monitoring, pumping, water transport, construction of all facilities, and all operation and maintenance, etc.) to city boundaries is on average 4.16 NIS/CM ($1.2/CM), whereas, all customers in the West Bank pay an average of only 2.85 NIS/CM ($0.8/CM) for receiving their water from Mekorot." [Mekorot, *Mekorot's Association with the Palestinians Regarding Water Supplies* (Tel Aviv: Mekorot, 2014), 13.]
3. Clive Lipchin, author's interview, New York, June 19, 2014.
4. Office of the United Nations Special Coordinator for the Middle East Peace Process (UNSCO), *Gaza in 2020: A Liveable Place?* (Jerusalem: UNSCO, August 2012), 12.
5. Shimon Tal, author's interview, Tel Aviv, October 18, 2013.

6. Haim Gvirtzman, *The Israeli-Palestinian Water Conflict: An Israeli Perspective* (Ramat Gan, Israel: The Begin-Sadat Center for Strategic Studies, January 2012), 3.
7. Ibid., 2–3.
8. Central Bureau of Statistics, *Census of Population 1967: West Bank of the Jordan, Gaza Strip and Northern Sinai Golan Heights* (Jerusalem: Central Bureau of Statistics), ix.
9. Gvirtzman, *op. cit.*, 3
10. Gidon Bromberg, e-mail to the author, March 13, 2015.
11. Ephraim Sneh, a former head of Israel's civil administration in the West Bank and later twice the Israeli deputy minister of defense, says that water governance in the West Bank was initially only focused on providing higher quality water and in greater supply, but that for a period during the late 1970s and early 1980s when settlement activity grew robustly, there was an interest in taking Palestinian water. He says that this attempt was often, but not always, frustrated by political opposition, media reports, and changes in national government control. In any event, to the extent that Israel had been interested in taking Palestinian water, he says that the internal conversations ended after the West Bank was linked to Israel's domestic water grid.* Alternatively, Clemens Messerschmidt, a German national pursuing a doctorate in hydrogeology who lives in the West Bank, believes that Israel's primary interest in its occupation of the West Bank is not security but as a means of controlling and using Palestinian water resources.**

    * Ephraim Sneh, author's interview, telephone, June 20, 2014.

    ** Clemens Messerschmidt, author's interview, telephone, July 9, 2014.
12. Palestinian Water Authority, *Annual Status Report on Water Resources, Water Supply, and Wastewater in the Occupied State of Palestine—2011* (Ramallah: Palestinian Water Authority, 2012), 44. For the purposes of this chapter, official Palestinian population statistics are accepted at face value despite disagreement as to their validity. Studies contest the Palestinian Central Bureau of Statistics' population figures claiming they are inflated both to obtain more aid and to project a lower per capita standard of living. [Bennett Zimmerman, Roberta Seid, and Michael L. Wise, *The Million Person Gap: The Arab Population in the West Bank and Gaza* (Ramat Gan, Israel: The Begin-Sadat Center for Strategic Studies, 2005).]
13. According to the Palestinian Water Authority, "West Bank water supply relies heavily on the import of water from Mekorot [Israel's national water utility]," with that source comprising more than fifty-five percent of the West Bank's domestic water. [Palestinian Water Authority, *op. cit.*, 28–32.]
14. Alon Tal, interview, *op. cit.*

15. According to the Embassy of the United States in Tel Aviv, in June 2008, "the tension between Hamas and Fatah within the PA started to register in the Palestine Water Authority (PWA), which until now has resisted politicization." [Embassy of the United States Tel Aviv, Israel, "Trilateral Water Meeting: Planning to Meet Scarcity," *Wikileaks*, 08TELAVIV1400 (June 30, 2008).]
16. Almotaz Abadi, author's interview, Ramallah, January 9, 2014.
17. Israeli-Palestinian Interim Agreement on the West Bank and the Gaza Strip (Washington, DC, September 28, 1995), Annex III—Protocol Concerning Civil Affairs, Appendix 1— Powers and Responsibilities for Civil Affairs, Article 40 —Water and Sewage.
18. Bashar Masri, author's interview, telephone, March 4, 2015.
19. Anne-Marie O'Connor and William Booth, "Israel to Let Water Flow to West Bank Development at Center of Political Feud," *The Washington Post*, February 28, 2015.
20. When later given the opportunity to confirm this quotation from what had been an "on the record" interview, the interview subject acknowledged saying it, but added a qualification not in the original interview that he was only referring to the period from 1993 to 1995 and technical discussions being conducted then. He also asked that even these qualified comments not be attributed to him by name. In that same original interview, the Palestinian water person also said: "I am still convinced that working together at the technical level could ease finding proper solutions if politics from both sides provide a supporting environment."
21. Alon Tal, author's interview, telephone, September 19, 2013.
22. Abadi, *op. cit.*
23. UNSCO, *op. cit.*, 11–12.
24. Yousef Abu Mayla, author's interview, telephone, September 16, 2013.
25. Ahmad Al-Yaqubi, *Sustainable Water Resources Management of Gaza Coastal Aquifer* (Presentation, Second International Conference on Water Resources and Arid Environment, 2006), 2.
26. The Palestinian Central Bureau of Statistics estimates that Gaza City's population in 2014 is 606,749 and that the population for the entirety of Gaza is 1,760,037. ["Estimated Population in the Palestinian Territory Mid-Year by Governorate, 1997–2016," Palestinian Central Bureau of Statistics, accessed on March 23, 2015: www.pcbs.gov.ps/Portals/_Rainbow/Documents/gover_e.htm.]
27. Yousef Abu Mayla, e-mail to the author, May 30, 2014.
28. See: note 4.

29. Fadel Kawash, author's interview, telephone, December 22, 2013.
30. Israel Central Bureau of Statistics, *op. cit.*, ix.
31. Palestinian Central Bureau of Statistics, *op. cit.*
32. UNSCO, *op. cit.,* 8.
33. Abdelrahman Tamimi, author's interview, Ramallah, January 9, 2013.
34. UNSCO, *op. cit.,* 11–12.
35. Israel continued to provide Gaza with 1.3 billion gallons of water per year after it withdrew its settlements in 2005. In 2015, Israel announced that it would double its water supply to 2.6 billion gallons per year. [Tovah Lazaroff, Sharon Udasin, and Yaakov Lappin, "Israel Helps Relieve Water Crisis in Gaza Strip by Doubling Supply," *The Jerusalem Post*, March 3, 2015.]
36. Kawash, *op. cit.*
37. Zvi Herman, author's interview, telephone, August 5, 2014.
38. Ibid.
39. Shannon McCarthy, e-mail to the author, October 15, 2014.
40. Nadav Cohen, author's interview, Tel Aviv, October 20, 2013.
41. Ibid.
42. In addition to the CINADCO and MEDRC programs, there were other valuable programs that enhanced Israeli-Palestinian and Israeli-Palestinian-Jordanian water cooperation that space does not allow discussion of in the main text. The EXACT program helped to create a database that was produced by Israeli, Palestinian, and Jordanian water professionals for joint use. MERC helped to facilitate Israeli and Palestinian dialog on key water problems. And the Red Sea–Dead Sea Conveyance Project brought Israeli, Jordanian, and Palestinian water professionals together.
43. Avi Aharoni, e-mail to the author, July 7, 2014.
44. Olga Slepner, e-mail to the author, November 27, 2014.
45. World Bank, Red Sea–Dead Sea Water Conveyance Study, *Draft Feasibility Study Report*, (Washington, DC: World Bank, July 2012), 12.
46. Uri Shani, author's interview, New York, December 10, 2013.
47. Some environmentalists have expressed a concern that adding a great amount of brine to the Dead Sea could result in one of many—still theoretical—environmental problems. The new water could evaporate more quickly than assumed and create a moist microclimate with unintended consequences. Or the different density of the brine might not mix with the water in the Dead Sea and create a stratified lake, also with unknown effect. Another fear expressed is tied to what will happen once the brine discharge is at full force many years from now, with a particular worry that the

newly added minerals in the water could, over time, turn the top of the lake white. And if not white, others fear that the new less salty Dead Sea might host algae blooms and turn the Dead Sea red or green. [Julia Amalia Heyer and Samiha Shafy, "Dead Sea: Environmentalists Question Pipeline Rescue Plan," *SpiegelOnline*, December 19, 2013.]

48. Shani, *op. cit.*
49. Ibid.
50. Seth M. Siegel, "A Middle East Accord—No Diplomats Needed," *The Wall Street Journal*, January 6, 2014.
51. Shani, *op. cit.*
52. Alon Tal and Alfred Abed Rabbo, *Water Wisdom: Preparing the Groundwork for Cooperative and Sustainable Water Management in the Middle East* (New Brunswick, NJ: Rutgers University Press, 2010).
53. Alfred Abed Rabbo, author's interview, telephone, October 5, 2014.
54. Adar, *op cit.* Professor Adar's comment may be building on an important work on pricing water to aid in conflict resolution. See: Franklin M. Fisher and Annette Huber-Lee, *Liquid Assets: An Economic Approach for Water Management and Conflict Resolution in the Middle East and Beyond* (Washington, DC: Resources for the Future, 2005).
55. Lipchin, *op. cit.*
56. Ibid.
57. While the Palestinians already have permits to build wastewater-treatment plants, few have been built. [Cohen, *op. cit.*]
58. Leila Hashweh, author's interview, telephone, July 2, 2014.
59. Although conditions described in the 2009 World Bank report have changed in many respects in the intervening years, the study highlights what still seems to be the constraints on development in Area C. [World Bank, *West Bank and Gaza: Assessment of Restrictions on Palestinian Water Sector,* Report No. 47657-GZ (Washington, DC: World Bank, April 2009), 34, 54, 55, 59, 135.]
60. Gidon Bromberg, author's interview, telephone, March 1, 2015.
61. Lipchin, *op. cit.*

CHAPTER 10: HYDRO-DIPLOMACY: ISRAEL'S USE OF WATER FOR GLOBAL ENGAGEMENT

1. Although the Soviet Union had briefly been supportive of Israel and voted for Israeli statehood at the UN, the USSR soon turned implacably hostile to Israel, seeing it as an

outpost of the West. For a history of USSR-Israel relations, and the USSR's policies in the Middle East, see: Galia Golan, *Soviet Policies in the Middle East: From World War Two to Gorbachev* (Cambridge: Cambridge University Press, 1990).

2. Zeev Shilo and Nissan Navo, *TAHAL: Chamishim Ha-Shanim Ha-Rishonim* [TAHAL: The First Fifty Years] (Israel: Shinar Publications, 2008), 241–242.
3. Ibid., 243.
4. Yosi Shalhevet, author's interviews, telephone, October 3 and 13, 2014. Dr. Shalhevet also generously shared a copy of the English translation of his memoir of his time in China. In English, it is called *China and Israel: Science in the Service of Diplomacy*.
5. Danny Tal, author's interview, telephone, October 22, 2014.
6. Huageng Pan, author's interview, New York, March 10, 2013.
7. Sharon Udasin, "Bennett Announces Water City for Israeli Technologies in Shougang, China," *The Jerusalem Post*, November 24, 2014.
8. Ninety-two percent of the freshwater withdrawal in Iran from 2000 to 2010 was by its agricultural sector. [Food and Agriculture Organization of the United Nations, *Food and Nutrition in Numbers 2014* (Rome: Food and Agriculture Organization of the United Nations, 2014), 48.]
9. Jeremy Sharp, *Water Scarcity in Iran: A Challenge for the Regime?* (Washington, DC: Library of Congress, Congressional Research Service, April 22, 2014).
10. Sediqeh Babran and Nazli Honarbakhsh, "Bohran Vaziat-e Ab Dar Jahan va Iran" [Water Crisis in the World and in Iran] (in Farsi), *Rahbord*, no. 48 (2008).
11. Masoud Tajrishy, "National Report of Iran," in *Mid-Term Proceedings on Capacity Development for the Safe Use of Wastewater in Agriculture*, eds. Reza Ardakanian, Hani Sewilam, Jens Liebe (Bonn, Germany: UN–Water Decade Program on Capacity Development, August 2012), 123.
12. Masoud Tajrishy and Ahmad Abrishamchi, "Integrated Approach to Water and Wastewater Management for Tehran, Iran," in *Water Conservation, Reuse, and Recycling: Proceedings of an Iranian-American Workshop*, ed. National Research Council (Washington, DC: The National Academies Press, 2005), 224.
13. Shmuel Aberbach, author's interview, telephone, March 10, 2014.
14. Arie Issar, author's interview, Jerusalem, April 24, 2013.
15. Arie Lova Eliav, *Letter to The New York Times*, March 1, 1979.
16. Judith A. Brown, "The Earthquake Disaster in Western Iran, September 1962," *Geography* 48, no. 2 (April 1963): 184–185.
17. Howard A. Patten, *Israel and the Cold War: Diplomacy, Strategy and the Policy of the Periphery at the United Nations* (New York: I. B. Tauris, 2013), 42.
18. Ibid., 43.

19. Alex Weisberg, author's interview, telephone, April 18, 2014.
20. Aberbach, *op. cit.*
21. Moshe Gablinger, author's interview, telephone, April 16, 2014.
22. Issar, *op. cit.* Arie Issar had several tours of duty inside Iran helping to build and expand its water system. After his first lengthy stay helping to rebuild the Qazvin water system after a devastating earthquake, he received a warm letter from Dr. Iraj Vahidi, Iran's deputy minister of water and power. The July 28, 1965, letter reads: "The Ministry of Water and Power has the honour to appreciate your precious services on [sic] the field of hydrogeology throughout your stay in Iran. Your close and heartily [sic] cooperation which apparently effects [sic] every Iranian individual will be never forgotten. On the occasion of your departure, I have the pleasure to offer you a humble present for remembrances of your stay in Iran." The present was a Turkoman rug. [Iraj Vahidi, *Letter to Arie Issar*, July 28, 1965.]
23. Gablinger, *op. cit.*
24. Issar, interview, *op. cit.*
25. Uri Lubrani, author's interview, telephone, May 4, 2014.
26. Moshe Gablinger, e-mail to the author, April 17, 2014.
27. Patten, *op. cit.*, 42–43.
28. Nathan Berkman, "Back in the Old Days," *Israel Desalination Society*, 2007.
29. IDE Technologies, *Reference List* (Tel Aviv: IDE Technologies, 2013).
30. Fredi Lokiec, author's interview, Kadima (Israel), May 1, 2013.
31. Yehuda Avner, *The Prime Ministers: An Intimate Narrative of Israeli Leadership* (New Milford, CT: Toby Press, 2010), 104–107.
32. Theodor Herzl, *Old New Land (Altneuland)* (Minneapolis, MN: Filiquarian Publishing, 2007), 193.
33. Avner, *op. cit.*, 105.
34. Haim Divon, author's interview, telephone, June 25, 2014.
35. Yehuda Avner, author's interview, telephone, March 19, 2013.
36. Israel Ministry of Foreign Affairs, MASHAV—Israel's Agency for International Development Cooperation, *Annual Report 2013* (Jerusalem: MASHAV), 18–23.
37. "About MASHAV," MASHAV—Israel's Agency for International Development Cooperation, accessed on March 24, 2015: mfa.gov.il/MFA/mashav/AboutMASHAV/Pages/Background.aspx.
38. Divon, *op. cit.*
39. Netafim and the other Israeli drip-irrigation companies have significantly changed the lives of poor farmers all around the world, and most especially in India. But TAHAL has designed and built the infrastructure that permits water to flow through the drip-irrigation system in many of those countries.

40. Paul H. Doron, *Seldom a Dull Moment: Memoirs of an Israeli Water Engineer* (Tel Aviv: Paul H. Doron, 1987), 202–414.
41. Joshua Schwarz, e-mail to the author, November 9, 2014.
42. Saar Bracha, author's interview, telephone, October 5, 2013.
43. Ibid.
44. India is Israel's largest buyer of military equipment. In October 2014, India completed a $520 million deal to buy Israeli missiles. In the first nine months of 2014, bilateral trade was a record $3.4 billion. [Tova Cohen and Ari Rabinovitch, "Under Modi, Israel and India Forge Deeper Business Ties," *Reuters*, November 23, 2014.]
45. TAHAL's assignment was to create a master plan for the water sources in Rajasthan. [Shilo and Navo, *op. cit.*, 244.]
46. Ibid, 244–248.
47. According to MVV's Web site, the "MVV Water Utility Pvt Ltd. is a consortium of SPML Infra, Tahal Consulting Engineers and Israel's largest [municipal] water company Hagihon Jerusalem Water and Wastewater Works formed to undertake the improvement in service level for water supply in Mehrauli and Vasant Vihar project areas." ["About Us," MVV Water Utility, accessed on February 7, 2015: mvvwater.com/about-us.html.]
48. Alon Yegnes, author's interview, telephone, November 5, 2014.
49. Moshe Gablinger, author's interview, telephone, October 23, 2014.
50. Sivan Ya'ari, author's interview, telephone, October 19, 2014.
51. Ruhakana Rugunda, author's interview, telephone, October 24, 2014.
52. Ya'ari, *op. cit.*
53. Meir Ya'acoby, author's interview, telephone, October 20, 2014.
54. Ya'ari, *op. cit.*

CHAPTER 11: NO ONE IS IMMUNE: CALIFORNIA AND THE BURDEN OF AFFLUENCE

1. Caroline Stauffer, "Election-Year Water Crisis Taking a Toll on Brazil's Economy," *Reuters*, October 31, 2014.
2. Luciana Magalhaes, Reed Johnson, and Paul Kiernan, "Blackouts Roll through Large Swath of Brazil," *The Wall Street Journal*, January 19, 2015.
3. Claire Rigby, "Sao Paulo—Anatomy of a Failing Megacity: Residents Struggle as Water Taps Run Dry," *The Guardian*, February 25, 2015.
4. Excluding environmental use, eighty percent of California's water is used for agriculture, a higher percentage than is common in the OECD countries. [Jeff Guo, "Agriculture is 80 percent of water use in California. Why aren't farmers being forced to cut back?" *The Washington Post*, April 3, 2015.]

5. Dan Keppen, author's interview, telephone, June 4, 2013.
6. Hillel Koren, "California, Israel to Join on Renewable Energy," *Globes*, November 15, 2009.
7. Governor Edmund G. Brown Jr. proclaimed a drought State of Emergency on January 17, 2014. ["Governor Brown Declares Drought State of Emergency," The Office of Governor Edmund G. Brown Jr., accessed on November 25, 2014: gov.ca.gov/news.php?id=18379.]
8. California-Israel Cooperation Agreement (Mountain View, CA, March 5, 2014).
9. Edmund G. Brown, *Speech after Signing the California-Israel Cooperation Agreement* (Mountain View, CA, March 5, 2014).
10. Benjamin Netanyahu, *Speech after Signing the California-Israel Cooperation Agreement* (Mountain View, CA, March 5, 2014).
11. Glenn Yago, author's interview, telephone, October 23, 2010.
12. Kish Rajan, author's interview, telephone, November 25, 2014.
13. Rebecca Salinas, "Texas Drought Will Lighten Up by Winter, Report Says," *My San Antonio*, August 22, 2014.
14. Udi Zuckerman, author's interview, Tel Aviv, January 6, 2014.
15. Texas Comptroller of Public Accounts, *Texas Water Report: Going Deeper for the Solution* (Austin, TX: Texas Comptroller of Public Accounts, 2014).
16. Rick Perry, author's interview, Tel Aviv, October 22, 2013.
17. "Another Warm Winter Likely for Western U.S., South May See Colder Weather," National Oceanic and Atmospheric Administration, accessed on November 25, 2014: www.noaanews.noaa.gov/stories2014/20141016_winteroutlook.html.
18. U.S. Government Accountability Office, *Freshwater: Supply Concerns Continue, and Uncertainties Complicate Planning*, GAO-14-430 (Washington, DC: U.S. Government Accountability Office, May 2014), 28.
19. Gwen N. Tindula, Morteza N. Orang, and Richard L. Snyder, "Survey of Irrigation Methods in California in 2010," *Journal of Irrigation and Drainage Engineering* 139, no. 3 (August 2013): 237.
20. U.S. Environmental Protection Agency, *2012 Guidelines for Water Reuse*, EPA/600/R-12/618 (Washington, DC: U.S. Environmental Protection Agency, September 2012), 5-1.

CHAPTER 12: GUIDING PHILOSOPHY

1. Shimon Peres, author's interview, Tel Aviv, April 25, 2013.
2. Haim Gvirtzman, author's interview, Tel Aviv, October 23, 2013.
3. Uri Shani, author's interview, telephone, March 17, 2013.

4. Gilad Fernandes, author's interview, Tel Aviv, October 28, 2013.
5. Ibid.
6. Ronen Wolfman, author's interview, October 24, 2013.
7. Yossi Shmaya, author's interview, Beit Netofa Valley (Israel), April 30, 2013.
8. Ori Yogev, author's interview, telephone, March 19, 2013.
9. If there is still a recognized weakness in the effort to depoliticize the national and local water systems, it is that a political figure, a Cabinet minister, still decides who heads the Water Authority, and the mayors still decide who sits on the municipal water utility board. In both cases, there is the potential, in theory if not in practice, for political intervention and favoritism. At least as of now, the Water Authority and the many municipal utility corporations are generally well-functioning, forward-looking, merit-based civil service organizations.
10. Shimon Tal, author's interview, Tel Aviv, January 6, 2014.
11. Nir Barlev, author's interview, telephone, April 11, 2013.
12. Yossi Yaacoby, author's interview, Tel Aviv, May 6, 2013.
13. Yossi Smoler, author's interview, telephone, March 18, 2014.
14. Yaacoby, *op. cit.*
15. Zohar Yinon, author's interview, Jerusalem, April 24, 2013.
16. Menachem Priel, author's interview, Tel Aviv, May 6, 2013.
17. Water Measurement Law, 5715-1955.
18. The well-travelled American water and soil expert Walter Clay Lowdermilk, the author of the 1944 book *Palestine, Land of Promise* profiled in Chapter 2, wrote in the late 1960s that: "The State of Israel made an inventory of all her land and water resources that is more thorough and complete than any other country I know." [Walter Clay Lowdermilk, "Water for the New Israel," late 1960s.]
19. In addition to painstakingly monitoring usage patterns, Israel has repeatedly attempted to determine the quantity of available natural water, whether from rainfall, Israel's aquifers, or other sources. This planning tool was given new emphasis by Professor Uri Shani when he was the head of the Israel Water Authority. By determining that Israel's natural sources of water were less than earlier assumed, it gave more urgency to the need to pursue man-made alternatives to Israel's natural water. [Shani, *op. cit.*]
20. Diego Berger, e-mail to the author, April 30, 2013.
21. Barlev, *op. cit.*
22. Shmaya, *op. cit.*
23. Berger, *op. cit.*

24. Michael Zaide, author's interview, Tel Aviv, April 25, 2013.
25. Priel, *op. cit.*
26. Pat Mulroy, author's interview, telephone, July 15, 2013.
27. Abraham Tenne, author's interview, Tel Aviv, April 25, 2013.

Selected Bibliography

BOOKS

Almogi, Yosef. *Total Commitment*. East Brunswick, NJ: Cornwall Books, 1982.

Amir, Giora. *Movil Ha-Mayim: Hayav U'Po'alo shel Aharon Wiener* [The Water Mover: The Life and Work of Aaron Wiener]. Kibbutz Daliya, Israel: Ma'arechet Publishing, 2012.

Avineri, Shlomo. *Herzl: Theodor Herzl and the Foundation of the Jewish State*. Translated by Haim Watzman. London: Weidenfeld & Nicolson, 2013.

Avner, Yehuda. *The Prime Ministers: An Intimate Narrative of Israeli Leadership*. New Milford, CT: The Toby Press, 2010.

Becker, Nir. *Water Policy in Israel: Context, Issues and Options*. Dordrecht, Netherlands: Springer, 2013.

Black, Edwin. *The Transfer Agreement: The Dramatic Story of the Pact Between the Third Reich and Jewish Palestine*. Northampton, MA: Brookline Books, 1999.

Blass, Simcha. *Mei Meriva u-Ma'as* [Water in Strife and Action]. Israel: Massada Ltd., 1973.

Doron, Paul H. *Seldom a Dull Moment: Memoirs of an Israeli Water Engineer.* Tel Aviv: Paul H. Doron, 1987.

Drinan, Joanne E. *Water and Wastewater Treatment: A Guide for the Nonengineering Professional.* Boca Raton, FL: CRC Press, 2001.

Elon, Amos. *Herzl*. New York: Holt, Rinehart and Winston, 1975.

Fisher, Franklin M., and Annette Huber-Lee. *Liquid Assets: An Economic Approach for Water Management and Conflict Resolution in the Middle East and Beyond.* Washington, DC: Resources for the Future, 2005.

Friling, Tuvia. *Arrows in the Dark*. Translated by Ora Cummings. Madison, WI: University of Wisconsin Press, 2005.

Gavron, Assaf. *Hidromanyah* [Hydromania]. Or Yehuda, Israel: Zmora-Bitan, Dvir Publishing House Ltd., 2008.

Gavron, Daniel. *The Kibbutz: Awakening from Utopia*. Lanham, MD: Rowman & Littlefield, 2000.

Gidor, Fanny. *Socio-Economic Disparities in Israel*. Piscataway, NJ: Transaction Publishers, 1979.

Gilbert, Martin. *In Ishmael's House: A History of Jews in Muslim Lands*. New Haven, CT: Yale University Press, 2010.

Golan, Galia. *Soviet Policies in the Middle East: From World War Two to Gorbachev*. Cambridge: Cambridge University Press, 1990.

Goldberg, Dan, Baruch Gornat, and Daniel Rimon. *Drip Irrigation: Principles, Design, and Agricultural Practices*. Kfar Shmaryahu, Israel: Drip Irrigation Scientific Publications, 1976.

Gordis, Daniel. *Menachem Begin: The Battle for Israel's Soul*. New York: Schocken, 2014.

Gvirtzman, Haim. *Mash'avei Ha-Mayim Be-Yisrael: Perakim Be-hydrologia U've Mada'ei Ha-Sevivah* [Israel Water Resources: Chapters in Hydrology and Environmental Sciences]. Jerusalem: Yad Ben-Zvi Press, 2002.

Hacohen, Dvora. *Immigrants in Turmoil: Mass Immigration to Israel and Its Repercussions in the 1950s and After*. Syracuse, NY: Syracuse University Press, 2003.

Herzl, Theodor. *Old New Land (Altneuland)*. Minneapolis, MN: Filiquarian Publishing, 2007.

Imeson, Anton. *Desertification, Land Degradation, and Sustainability*. Hoboken, NJ: Wiley, 2012.

Kally, Elisha, and Gideon Fishelson. *Water and Peace: Water Resources and the Arab-Israeli Peace Process*. Westport, CT: Praeger, 1993.

Kellerman, Aharon. *Society and Settlement: Jewish Land of Israel in the Twentieth Century*. Albany, NY: State University of New York Press, 1993.

Keren, Zwi. *Oasis in the Desert: The Story of Kibbutz Hatzerim*. Translated by Kfar Blum. Kibbutz Hatzerim, Israel: Kibbutz Hatzerim Press, 1988.

Lowdermilk, Inez Marks. *All in a Lifetime: An Autobiography*. Berkeley, CA: The Lowdermilk Trust, 1985.

Lowdermilk, Walter Clay. *Palestine, Land of Promise*. 3rd ed. New York: Harper & Brothers, 1944.

Makovsky, Michael. *Churchill's Promised Land: Zionism and Statecraft*. New Haven, CT: Yale University Press, 2008.

McCarthy, Justin. *The Population of Palestine: Population History and Statistics of the Late Ottoman Period and the Mandate*. New York: Columbia University Press, 1990.

Mekorot. *60 Shanah Le-Kav Ha-Rishon La-Negev* [60 Years to the First Pipeline to the Negev]. Tel Aviv: Mekorot, 2007.

Metzer, Jacob. *The Divided Economy of Mandatory Palestine*. Cambridge: Cambridge University Press, 1998.

Mithen, Steven. *Thirst: Water and Power in the Ancient World*. Cambridge, MA: Harvard University Press, 2012.

Morris, Benny. *Righteous Victims: A History of the Zionist-Arab Conflict, 1881–2001*. New York: Vintage, 2001.

Morris, Yaakov. *Masters of the Desert: 6,000 Years in the Negev*. New York: G. P. Putnam's Sons, 1961.

Oz, Amos. *My Michael*. New York: Knopf, 1972.

Palgi, Michael, and Shulamit Reinharz. *One Hundred Years of Kibbutz Life: A Century of Crises and Reinvention*. New Brunswick, NJ: Transaction, 2011.

Patten, Howard A. *Israel and the Cold War: Diplomacy, Strategy and the Policy of the Periphery at the United Nations*. New York: I. B. Tauris, 2013.

Postel, Sandra. *Pillar of Sand: Can the Irrigation Miracle Last?* New York: W. W. Norton & Company, 1999.

Rayman, Paula. *The Kibbutz Community and Nation Building*. Princeton, NJ: Princeton University Press, 1981.

Rivlin, Paul. *The Israeli Economy from the Foundation of the State through the 21st Century*. Cambridge: Cambridge University Press, 2011.

Rubin, Barry, and Wolfgang G. Schwanitz. *Nazis, Islamists, and the Making of the Modern Middle East*. New Haven, CT: Yale University Press, 2014.

Sachar, Howard M. *A History of Israel: From the Rise of Zionism to Our Time*. New York: Knopf, 1976.

Salzman, James. *Drinking Water: A History*. New York: Overlook Duckworth, 2012.

Segev, Tom. *The Seventh Million: The Israelis and the Holocaust*. Translated by Haim Watzman. New York: Hill & Wang, 1993.

Seltzer, Assaf. *Meḳorot: Sipurah Shel Hevrat Ha-Mayim Ha-Le'umit—75 Ha-Shanim Ha-Rishonot* [Mekorot: The Story of the Israel National Water Company—The First 75 Years]. Jerusalem: Yad Yitzhak Ben-Zvi, 2011.

Senor, Dan, and Saul Singer. *Start-Up Nation: The Story of Israel's Economic Miracle*. New York: Twelve, 2009.

Shalhevet, Joseph. *China and Israel: Science in the Service of Diplomacy*. Israel: Joseph Shalhevet, 2009.

Shilo, Zeev, and Nissan Navo. *TAHAL: Chamishim Ha-Shanim Ha-Rishonim* [TAHAL: The First Fifty Years]. Israel: Shinar Publications, 2008.

Shoham, Yael, and Ofra Sarig. *Ha-Movil Ha-Artzi: Min Ha-Kineret ve-ad Pe-atei Negev* [The National Water Carrier: From the Sea of Galilee to the Negev]. Tel Aviv: Mekorot, 1995.

Smith, Charles D. *Palestine and the Arab-Israeli Conflict*. 6th ed. Boston: Bedford/St. Martin's, 2007.

Soffer, Arnon. *Rivers of Fire: The Conflict over Water in the Middle East*. Translated by Murray Rosovsky and Nina Copaken. Lanham, MA: Rowman & Littlefield, 1999.

Sosland, Jeffrey. *Cooperating Rivals: The Riparian Politics of the Jordan River Basin*. Albany, NY: State University of New York Press, 2007.

Solomon, Steven. *Water: The Epic Struggle for Wealth, Power and Civilization*. New York: Harper Perennial, 2010.

Strong, James. *Strong's Exhaustive Concordance of the Bible*. Peabody, MA: Hendrickson Publishers, 2009.

Tal, Alon. *Pollution in a Promised Land: An Environmental History of Israel.* Berkeley, CA: University of California Press, 2002.

Tal, Alon, and Alfred Abed Rabbo. *Water Wisdom: Preparing the Groundwork for Cooperative and Sustainable Water Management in the Middle East.* New Brunswick, NJ: Rutgers University Press, 2010.

Wiener, Aaron. *Development and Management of Water Supplies Under Conditions of Scarcity of Resources.* Tel Aviv: TAHAL, April 1964.

Wolf, Aaron. *Hydropolitics along the Jordan River: Scarce Water and its Impact on the Arab-Israeli Conflict*. Tokyo: United Nations University Press, 1995.

Yehoshua, Abraham B. *Early in the Summer of 1970.* New York: Schocken, 1971.

BOOK CHAPTERS

Alatout, Samer. "Hydro-Imaginaries and the Construction of the Political Geography of the Jordan River: The Johnston Mission, 1953–56." In *Environmental Imaginaries of the Middle East and North Africa*. Edited by Diana K. Davis and Edmund Burke III. Athens, OH: Ohio University Press, 2011.

Dowty, Alan. "Israel's First Decade: Building a Civic State." In *Israel: The First Decade of Independence*. Edited by Selwyn K. Troen and Noah Lucas. Albany: State University of New York Press, 1995.

Gleick, Peter H., and Matthew Heberger. "Water and Conflict: Events, Trends, and Analysis (2011–2012)." In *The World's Water*, vol. 8. Edited by Peter H. Gleick. Oakland, CA: Pacific Institute for Studies in Development, Environment and Security, 2014.

Gordon, A. D. "Our Tasks Ahead." In *The Zionist Idea: A Historical Analysis and Reader*. Edited by Arthur Hertzberg. Philadelphia: Jewish Publication Society, 1997.

Kislev, Yoav. "Agricultural Cooperatives in Israel, Past and Present." In *Agricultural Transition in Post-Soviet Europe and Central Asia after 20 Years*. Edited by A. Kimhi and Z. Lerman. Halle, Germany: Leibniz Institute of Agricultural Development in Transition Economies.

Laster, Richard, and Dan Livney. "Israel: The Evolution of Water Law and Policy." In *The Evolution of the Law and Politics of Water*. Edited by Joseph W. Dellapenna and Joyeeta Gupta. Dordrecht, Netherlands: Springer, 2009.

Lowdermilk, Walter Clay. "Israel: A Pilot Project for Total Development of Water Resources." In *Essays in Honor of Abba Hillel Silver*. Edited by Daniel Jeremy Silver. New York: MacMillan, 1963.

Tajrishy, Masoud. "National Report of Iran." In *Mid-Term Proceedings on Capacity Development for the Safe Use of Wastewater in Agriculture*. Edited by Reza Ardakanian, Hani Sewilam, and Jens Liebe. Bonn, Germany: UN–Water Decade Program on Capacity Development, August 2012.

Tajrishy, Masoud, and Ahmad Abrishamchi. "Integrated Approach to Water and Wastewater Management for Tehran, Iran." In *Water Conservation, Reuse, and Recycling: Proceedings of an Iranian-American Workshop*. Edited by National Research Council. Washington, DC: The National Academies Press, 2005.

JOURNALS, EXTENDED ARTICLES, AND PH.D. THESES

Abrahams, Harold J. "The Hezekiah Tunnel." *Journal–American Water Works Association* 70, no. 8 (August 1978): 406–410.

Al-Ansari, Nadhir, N. Alibrahiem, M. Alsaman, and Sven Knutsson. "Water Supply Network Losses in Jordan." *Journal of Water Resource and Protection* 6, no. 2 (February 2014): 83–96.

Alatout, Samer. "'States' of Scarcity: Water, Space, and Identity Politics in Israel, 1948–59." *Environmental Planning D: Society and Space* 26, no. 6 (July 2008): 959–982.

Alqadi, Khaled A., and Lalit Kumar. "Water Policy in Jordan." *International Journal of Water Resources Development* 30, no. 2 (2014): 322–334.

Babran, Sediqeh, and Nazli Honarbakhsh. "Bohran Vaziat-e Ab Dar Jahan va Iran" [Water Crisis in the World and in Iran] (In Farsi). *Rahbord*, no. 48 (2008).

Bartlett, Bruce. "The Crisis of Socialism in Israel." *Orbis* 35 (1991): 53–61.

Berkman, Nathan. "Back in the Old Days." *Israel Desalination Society*, 2007.

Birkett, James D. "A Brief Illustrated History of Desalination: From the Bible to 1940." *Desalination* 50 (1984): 17–52.

Brown, Judith A. "The Earthquake in Western Iran, September 1962." *Geography* 48, no. 2 (April 1963): 184–185.

Femia, Francesco, and Caitlin Werrell. "Syria: Climate Change, Drought and Social Unrest." *The Center for Climate and Security*, March 3, 2012.

Gabay, Shoshana. "Restoring Israel's Rivers." *Israel Ministry of Foreign Affairs*, 2001.

Gronewold, Andrew D., and Craig A. Stow. "Water Loss from the Great Lakes." *Science* 343, no. 6175 (March 7, 2014): 1084–1085.

Heffez, Adam. "How Yemen Chewed Itself Dry." *Foreign Affairs*, July 23, 2013.

Herzog, Donna M. "Contested Waterscapes: Constructing Israeli Water and Identity." PhD Dissertation. New York University, 2015.

Johnson, Lyndon B. "If We Could Take the Salt Out of Water." *The New York Times Magazine*, October 30, 1960.

Kay, Avi. "From *Altneuland* to the New Promised Land: A Study of the Evolution and Americanization of the Israeli Economy." *Jewish Political Studies Review* 24, no. 1–2 (2012): 99–128.

Lowdermilk, Walter Clay. "Water for the New Israel." Unpublished, 1967/68.

Mane, Amir. "Americans in Haifa: The Lowdermilks and the American-Israeli Relationship." *Journal of Israeli History* 30, no. 1 (2011): 65–82.

Maurer, Noel, and Carlos Yu. "What Roosevelt Took: The Economic Impact of the Panama Canal, 1903–37." *Harvard Business School*, 06-041 (March 2006).

Morag, Nadav. "Water, Geopolitics and State Building: The Case of Israel." *Middle Eastern Studies* 37, no. 3 (2001): 179–198.

Ofek, Elie, and Matthew Preble. "TaKaDu." Harvard Business School, Case Study, 514-083 (January 2014).

Postel, Sandra. "Growing More Food with Less Water." *Scientific American* 284, no. 2 (2001): 46–59.

Reichman, Shalom, Yossi Katz, and Yair Paz. "The Absorptive Capacity of Palestine, 1882–1948." *Middle Eastern Studies* 33, no. 2 (1997): 338–361.

Reig, Paul, Andrew Maddocks, and Francis Gassert. "*World's 36 Most Water-Stressed Countries.*" *World Resources Institute*, December 12, 2013.

Ron, Zvi Y. D. "Ancient and Modern Developments of Water Resources in the Holy Land and the Israeli-Arab Conflict: A Reply." *Transactions of the Institute of British Geographers, New Series* 11, no. 3 (1986): 360–369.

Sneddon, Chris, and Coleen Fox. "The Cold War, the US Bureau of Reclamation, and the Technopolitics of Tiver Basin Development, 1950–1970." *Political Geography* 30 (2011): 450–460.

Stemple, John. "Viewpoint: A Brief Review of Afforestation Efforts in Israel." *Rangelands* 20, no. 2 (April 1998): 15–18.

Tal, Alon. "Thirsting for Pragmatism: A Constructive Alternative to Amnesty International's Report on Palestinian Access to Water." *Israel Journal of Foreign Affairs* 4, no. 2 (2010): 59–73.

Tenne, Abraham, Daniel Hoffman, and Eytan Levi. "Quantifying the Actual Benefits of Large-Scale Seawater Desalination in Israel." *Desalination and Water Treatment* 51, no. 1–3 (July 2012): 26–37.

Tindula, Gwen N., Morteza N. Orang, and Richard L. Snyder. "Survey of Irrigation Methods in California in 2010." *Journal of Irrigation and Drainage Engineering* 139, no. 3 (August 2013): 233–238.

Voss, Katalyn A., James S. Famiglietti, MinHui Lo, Caroline de Linage, Matthew Rodell, and Sean C. Swenson. "Groundwater Depletion in the Middle East from GRACE with Implications for Transboundary Water Management in the Tigris-Euphrates-Western Iran Region." *Water Resources Research* 49, no. 2: 904–914.

"U.S., Israel Finally to Build Horizontal Tube Prototype at Ashdod." *Water Desalination Report* 11, no. 27 (July 3, 1975).

Wolfowitz, Paul. "Nuclear Proliferation in the Middle East: The Politics and Economics of Proposals for Nuclear Desalting." PhD Dissertation. University of Chicago, 1972.

Zwingle, Erla. "Ogallala Aquifer: Well Spring of the High Plains." *National Geographic* 183, no. 3: 80–109.

REPORTS, PAPERS, AND PRESENTATIONS

Al-Yaqubi, Ahmad. *Sustainable Water Resources Management of Gaza Coastal Aquifer.* Presentation, Second International Conference on Water Resources and Arid Environment, 2006.

Amnesty International. *Thirsting for Justice: Palestinian Access to Water Restricted.* London: Amnesty International Publications, 2009.

———. *Troubled Waters—Palestinians Denied Fair Access to Water.* London: Amnesty International Publications, 2009.

Anglo-American Committee of Inquiry. *A Survey of Palestine: Prepared in December 1945 and January 1946 for the Information of the Anglo-American Committee of Inquiry.* Jerusalem: Government Printer, 1946–47.

Arab Republic of Egypt, Ministry of Water and Irrigation. *Integrated Water Resources Management Plan.* Cairo: Ministry of Water and Irrigation, June 2005.

Attili, Shaddad. *Israel and Palestine: Legal and Policy Aspects of the Current and Future Joint Management of the Shared Water Resources.* Ramallah: Palestine Liberation Organization Negotiation Support Unit, 2004.

Blass, Simcha. *Drip Irrigation.* Tel Aviv: Water Works—Consulting and Design, July 1969.

Brooks, David B., and Julie Trottier. *An Agreement to Share Water between Israelis and Palestinians: The FoEME Proposal.* Amman, Bethlehem, and Tel Aviv: Friends of the Earth Middle East, November 2010.

Central Bureau of Statistics. *Census of Population 1967: West Bank of the Jordan, Gaza Strip and Northern Sinai Golan Heights.* Jerusalem: Central Bureau of Statistics, 1967.

———. *Israel in Statistics: 1948–2007.* Jerusalem: Central Bureau of Statistics, May 2009.

City of Williams. *Level 4 Water Restrictions: URGENT NOTICE.* Williams, AZ: City of Williams, 2014.

Cotton, John S. *Plan for the Development of the Water Resources of the Jordan and Litani River Basins.* Jerusalem: Ministry of Agriculture, 1954.

Electric Power Research Institute. *Water and Sustainability (Volume 4): U.S. Electricity Consumption for Water Supply and Treatment—The Next Half Century.* Palo Alto, CA: Electric Power Research Institute, March 2002.

FAFO Research Foundation. *Iraqis in Jordan: Their Number and Characteristics.* Oslo: FAFO Research Foundation, May 2007.

Food and Agriculture Organization of the United Nations. *Statistical Yearbook 2013, World Food and Agriculture.* Rome: Food and Agriculture Organization of the United Nations, 2013.

Gafni, Wolf, Pinhas Michaeli, Ahouva Bar-Lev, Yerahmiel Barylka, and Edward Levin. *Beside Streams of Waters: Rain and Water in the Prayers and Ceremonies of the Holiday.* Jerusalem: Jewish National Fund, Keren Kayemeth LeYisrael, Religious Organizations Department, 1990.

Government of India, Ministry of Agriculture. *Report on the Task Force on Microirrigation.* New Delhi: Ministry of Agriculture, January 2004.

Government of Israel. *The Central Arava: Proposals for the Development of Water Resources,* Report 69-093. Jerusalem: Government of Israel, September 1969.

Gvirtzman, Haim. *The Israeli-Palestinian Water Conflict: An Israeli Perspective.* Ramat Gan, Israel: The Begin-Sadat Center for Strategic Studies, January 2012.

IDE Technologies Ltd. *IDE Technologies Limited.* Presentation, Gal Zohar, CFO Forum, 2011.

———. *Reference List.* Tel Aviv: IDE Technologies Ltd., 2013.

International Commission on Irrigation and Drainage. *World Irrigated Area-Region Wise/Country Wise.* New Delhi: International Commission on Irrigation and Drainage, 2012.

———. *Sprinkler and Micro Irrigated Areas.* New Delhi: International Commission on Irrigation and Drainage, May 2012.

International Fund for Agricultural Development. *Smallholders, Food Security and the Environment* (Rome: International Fund for Agricultural Development, 2013).

Israel Ministry of Foreign Affairs, MASHAV—Israel's Agency for International Development Cooperation, Annual Report 2013 (Jerusalem: Ministry of Foreign Affairs).

Israel State Comptroller. *Report for 2011*. Jerusalem: State Comptroller, December 2011.

Israel Water Authority. *Sea Water Desalination in Israel: Planning, Coping with Difficulties, and Economic Aspects of Long-Term Risks*. Edited by Abraham Tenne. Jerusalem: Israel Water Authority, October 2010.

———. *The Master Plan for Desalination in Israel, 2020*. Edited by Abraham Tenne. Jerusalem: Israel Water Authority, October 2011.

———. *The Water Issue between Israel and the Palestinians: Main Facts*. Jerusalem: Israel Water Authority, February 2012.

———. *The State Department 2012 Human Rights Report: Responses of the Water Authority to the Water Related Palestinian Arguments as Presented in the Report*. Jerusalem: Israel Water Authority, 2012.

———. *Economics Aspects in Water Management in Israel: Policy and Prices*. Edited by Gilad Fernandes. Jerusalem: Israel Water Authority, Date unknown.

Jordan River Rehabilitation Administration. *The Lower Jordan River: Rehabilitation and Landscape Development Master Plan*. Edited by Ram Aviram. Jerusalem: Jordan Rehabilitation Administration, January 2014.

King, Marcus DuBois. *Water, U.S. Foreign Policy and American Leadership*. Washington, DC: Elliott School of International Affairs, George Washington University, October 2013.

Kislev, Yoav. *The Water Economy of Israel*. Jerusalem: Taub Center for Social Policy Studies in Israel, November 2011.

Kharas, Homi, and Geoffrey Gertz. *The New Global Middle Class: A Cross-Over from West to East*. Washington, DC: Wolfensohn Center for Development at Brookings, 2010.

Lokiec, Fredi. *South Israel 100 Million m³/Year Seawater Desalination Facility: Build, Operate and Transfer (BOT) Project*. Kadima, Israel: IDE Technologies Ltd., March 2006.

Lowdermilk, Walter Clay. *Conquest of the Land through Seven Thousand Years*. Washington, DC: U.S. Department of Agriculture, 1948.

Lux Research. *Making Money in the Water Industry*, LRWI-R-13-4. New York: Lux Research, December 2013.

MacGowan, Charles F. *History, Function, and Program of the Office of Saline Water*. Presentation, New Mexico Water Conference, July 1–3, 1963.

Malik, Ravinder P. S., and M. S. Rathore. *Accelerating Adoption of Drip Irrigation in Madhya Pradesh, India, AgWater Solutions Project*. Sri Lanka: IWMI, September 2012.

MATIMOP: Israeli Industry Center for R&D. *International Cooperation and Government Support for R&D: The Israeli Case Study*. Ed. Michel Hivert. Jerusalem: MATIMOP, 2012.

Mekorot. *Masskinot Veh-ah-dat Ha-Yarkon* [Conclusions of the Committee on the Yarkon]. Edited by Simcha Blass. Tel Aviv: Mekorot, March 18, 1954.

———. *Wastewater Reclamation and Reuse*. Edited by Batya Yadin, Adam Kanarek, and Yael Shoham. Tel Aviv: Mekorot, 1993.

———. *Mekorot's Association with the Palestinians Regarding Water Supplies*. Tel Aviv: Mekorot, 2014.

Ministry of Agriculture and Rural Development. *Irrigation Agriculture—The Israeli Experience*. Edited by Anat Lowengart-Aycicegi. Jerusalem: Ministry of Agriculture.

———. *Technological Incubator's Program*. Presentation, Yossi Smoler, October 17, 2010.

Ministry of Economy, Office of the Chief Scientist. *Research and Development 2012–14*. Jerusalem: Office of the Chief Scientist, September 2014.

Netafim. *Irrigation & Strategies for Investment*. Presentation, Naty Barak, Agricultural Investment 2011, London, October 5–6, 2011.

———. *Drip Irrigation—Israeli Innovation That Has Changed the World*. Presentation, Naty Barak, JNF Summit, Las Vegas, April 28, 2013.

Organization for Economic Co-operation and Development (OECD). *Main Science and Technology Indicators Volume 2014*. Issue 2. Paris: OECD Publishing, 2015.

Office of the Director of National Intelligence, National Intelligence Council. *Global Water Security*, Intelligence Community Assessment. Washington, DC: National Intelligence Council, February 2, 2012.

———. *Global Trends 2025: A Transformed World*. Washington, DC: National Intelligence Council, November 2008.

Office of the United Nations Special Coordinator for the Middle East Peace Process. *Gaza in 2020: A Liveable Place?* Jerusalem: UNSCO, August 2012.

Palestinian Central Bureau of Statistics. *Palestine in Figures, 2004*. Ramallah: Palestinian Central Bureau of Statistics, May 2005.

———. *The Statistical Report, Household Environment Study*. Ramallah: Palestinian Central Bureau of Statistics, November 2013.

Palestinian Water Authority. *Palestinian Water Sector: Status Summary Report September 2012*. Ramallah: Palestinian Water Authority, September 2012.

———. *Annual Status Report on Water Resources, Water Supply, and Wastewater in the Occupied State of Palestine—2011*. Ramallah: Palestinian Water Authority, 2012.

———. *Water Sector Reform Plan 2014–16 (Final)*. Ramallah: Palestinian Water Authority, December 2013.

Parliamentary [Knesset] Commission of Inquiry with Regard to Lessons to Be Learned from the Maccabiah Bridge Disaster. *Report*. Jerusalem: Knesset: July 9, 2000.

Pullabhotla, Hemant K., Chandan Kumar, and Shilp Verma. *Micro-Irrigation Subsidies in Gujarat and Andhra Pradesh: Implications for Market Dynamics and Growth*. Sri Lanka: IWMI-TATA Water Policy Program, 2012.

Sharp, Jeremy. *Jordan: Background and U.S. Relationships*. Washington, DC: Library of Congress, Congressional Research Service, May 8, 2014.

———. *U.S. Foreign Aid to Israel*. Washington, DC: Library of Congress, Congressional Research Service, April 11, 2014.

———. *Water Scarcity in Iran: A Challenge for the Regime?* Washington, DC: Library of Congress, Congressional Research Service, April 22, 2014.

Smart Water Networks Forum Research. *Stated NRW (Non-Revenue Water) Rates in Urban Networks*. Portsmouth, U.K.: Smart Water Networks Forum, August 2010.

Shuval, Hillel I. *Public Health Aspects of Waste Water Utilization in Israel*. Presentation, Purdue Industrial Wastes Conference, May 1, 1962.

Swirski, Shlomo, and Yael Hasson. *Invisible Citizens: Israel Government Policy toward the Negev Bedouin*. Translated by Ruth Morris. Tel Aviv: Adva Center, 2006.

Texas Comptroller of Public Accounts. *Texas Water Report: Going Deeper for the Solution*. Austin, TX: Texas Comptroller of Public Accounts, 2014.

The Government of the Hashemite Kingdom of Jordan, Ministry of Water and Irrigation. *Red Sea–Dead Sea Project/Phase 1*. Amman: Ministry of Water and Irrigation, January 2014.

The Israel Export and International Cooperation Institute, *Israel's Agriculture*. Tel Aviv: The Israel Export and International Cooperation Institute, 2012.

The Parliamentary [Knesset] Committee of Inquiry on the Israeli Water Sector. *Report*. Jerusalem: Knesset, June 2002.

Tolts, Mark. *Post-Soviet Aliyah and Jewish Demographic Transformation*. Presentation, Fifteenth World Congress of Jewish Studies, 2009.

Trigger Consulting. *Seizing Israel's Opportunities in the Global Water Market*. Tel Aviv: Trigger Consulting, 2005.

United Nations World Water Assessment Program. *The United Nations World Water Development Report 2014: Water and Energy*. Paris: United Nations Education, Scientific and Cultural Organization, 2014.

U.S. Agency for International Development. *West Bank/Gaza: Water Resources and Infrastructure Program*. Washington, DC: U.S. Agency for International Development, January 2013.

U.S. Department of Interior, Office of Saline Water. *Office of Saline Water's Report (Condensed)*. Washington, DC: Office of Saline Water, 1962.

U.S. Government Accountability Office. *Freshwater: Supply Concerns Continue, and Uncertainties Complicate Planning*, GAO-14-430. Washington, DC: U.S. Government Accountability Office, May 2014.

World Bank, Red Sea–Dead Sea Water Conveyance Study Program. *Dead Sea Study: Draft Final Report*. Washington, DC: Red Sea–Dead Sea Water Conveyance Study Program, April 2011.

———. *Preliminary Environmental and Social Assessment*. Washington, DC: Red Sea–Dead Sea Water Conveyance Study Program, July 2012.

World Bank. *Water, Electricity, and the Poor: Who Benefits from Utility Subsidies?* Edited by Kristin Komives, Vivien Foster, Jonathan Halpern, and Quentin Wodon. Washington, DC: World Bank, 2005.

———. *West Bank and Gaza: Assessment of Restrictions on Palestinian Water Sector*, Report No. 47657-GZ. Washington, DC: World Bank, April 2009.

Zimmerman, Bennett, Roberta Seid, and Michael L. Wise. *The Million Person Gap: The Arab Population in the West Bank and Gaza*. Ramat Gan, Israel: The Begin-Sadat Center for Strategic Studies, 2005.

Index